# Technological Change Handbook

# Technological Change Handbook

Edited by **Ed Diego**

New York

Published by NY Research Press,
23 West, 55th Street, Suite 816,
New York, NY 10019, USA
www.nyresearchpress.com

**Technological Change Handbook**
Edited by Ed Diego

International Standard Book Number: 978-1-63238-435-5 (Hardback)

Printed in the United States of America.

# Contents

# Preface

The world is advancing at a fast pace like never before. Therefore, the need is to keep up with the latest developments. This book was an idea that came to fruition when the specialists in the area realized the need to coordinate together and document essential themes in the subject. That's when I was requested to be the editor. Editing this book has been an honour as it brings together diverse authors researching on different streams of the field. The book collates essential materials contributed by veterans in the area which can be utilized by students and researchers alike.

Technology is dynamic. Technological change, today, is central to economic growth. It is recognized as a significant driver of efficiency development and the appearance of new products from which customers benefit. It depends not only on the work of experts but also on a broader variety of financial and societal factors, inclusive of institutions such as academic property rights and commercial domination, the operation of markets, a range of governmental policies (science and technology policy, modernization policy, opposition policy, etc.) and other factors. Given that technology is openly taken up in the approaches and policies of administration, it is imperative that the nature and dynamics of technology be understood properly. This book intends to provide useful information regarding technological change to its readers.

Each chapter is a sole-standing publication that reflects each author's interpretation. Thus, the book displays a multi-facetted picture of our current understanding of application, resources and aspects of the field. I would like to thank the contributors of this book and my family for their endless support.

Editor

# Part 1

# Conceptualizing Technological Change

# Technological Change and Economic Transformation

Musa Jega Ibrahim
*Economic Research and Policy Department,*
*Islamic Development Bank, Jeddah*
*Kingdom of Saudi Arabia*

## 1. Introduction

Technological change is a term used to describe incremental change in the quality and quantity of knowledge and ideas that are applied in the stream of activities to enhance the social and economic well being of the society. Due to the positive nature of the implied change, it is also referred to as technological progress. Technological change occurs through the process of invention, innovation and diffusion that leads to the transformation of ideas and knowledge into tangible products that have high utility value to human needs. The effect of technological change propels economic transformation; a change in the structure of an economy over time from a lower, rudimentary and subsistence level to a higher and more sophisticated level of economic activities. Thus, economic transformation is the attainment of significant high level of economic growth above previous levels with capacity to sustain it through self-perpetuating economic activities that are associated with industrial and post-industrial production activities[1].

Economic transformation stems from high sustainable economic growth that feeds from, and into technological change. While the acquisition and application of technology is a key factor in achieving economic transformation, economic activities are in turn, veritable source of technological progress. Hence, economic growth, economic transformation and technological change are intervolving activities that reinforce each other. This three-dimensional relationship reflects in the definition of economic growth as a long term rise in capacity to supply increasingly diverse economic goods to the population based on advancing technology and the institutional and ideological adjustments that it demands (Kuznets, 1971).

Real productive activities engender economic growth by ensuring a continuous improvement in the methods of production, discovery of new resources and thus creating the necessary conditions for efficient utilisation of resources. Resources, in their natural form, have limited direct economic use in satisfying human needs but transforming them into goods and services enhances their economic value to the society. The process of transforming resources involve substantial mix of ideas (technology) with other factors of

---

[1] Some have ascribed ideological connotation into the term thus referring to economic transformation as transition from centrally planned economies towards open market economies.

production such as land and labour, in addition to other resources from different activity sectors of the economy. A multiple sector positive performance is essential for the growth of the overall economy, but a sector of the economy that attracts large spectrum of economic activities can stimulate the productive fibre of other sectors towards real production and provide the requisite impetus for sustainable growth of the economy. This requires the catalyst of technological application and thus underlining the essence of technological change as a critical determinant of economic growth and by extension, economic transformation.

In general, sources of technological change are innovation, direct acquisition from purchase, learning-by-doing, Research and Development (R&D) and transfer through interactions of economic activities between two countries (technology transfer). Experiences of economic development of countries indicate that, acquisition and application of technology depends largely on economic circumstances and natural endowments of countries, nevertheless it is imperative for all economies to adapt to technological change to inspire economic transformation that springs into high sustainable growth and prosperity.

In the remainder of this chapter, section 2 highlights the perspectives of economic growth theories on the relationship between technological change and economic growth. Section 3 analyzes the dimensions of technological change and economic transformation and section 4 articulate measures for fostering technological change and economic transformation.

## 2. Perspectives of economic growth theories

### 2.1 Overview

Economic growth occurs through the transformation of resources into different forms of use involving the interactions of variables such as demand, supply, wages and prices. Basically, economic growth is driven by a process that is generated and sustained by the efficient utilisation of economic resources to meet effective demand and social needs. Increase in the outputs of major sectors of an economy, especially manufacturing, due to increase in the use of inputs or improvement in technology, leads to economic growth. Progressive increase in the outputs of major sectors of an economy that stems mainly from efficient utilization of economic resources and through the effective use of technology leads to high and sustainable economic growth, a *sine qua non* for economic transformation.

From the theoretical literature, economic growth process is based on intricate interaction of variables relating to the basic components of the economic system. The basic ideas of modern growth theories are based on competitive behaviour and equilibrium dynamics, diminishing returns and its relation to the accumulation of physical and human capital, the interplay between per capita income and the growth rate of population, the effects of technological progress as increased specialization of labour and discoveries of new goods and methods of production and the role of market structure (monopoly and/or competition) as an incentive to technological advancement.

Economic growth theory has evolved over the years leading to different cluster of economic growth theories based on ascription to common principles in their strands of analysis of economic growth. Even though there are several of these cluster of economic growth theories, modern economic growth analysis are dominated by the neoclassical and endogenous, which are regarded as the two broad classifications of economic growth theory (McCallum, 1996). Each of these broad categories of theory has variants of economic analysis and they all converge on the critical role of technological change as the driving force of high and sustainable economic growth.

## 2.2 Neoclassical growth theory

This cluster of economic growth analysis was inspired by the earlier works of classical economists but the Solow and Ramsey-Cass-Koopmans (RCK) models have emerged as the most recognized neoclassical economic growth theories. They underlined the effects of technological change on increased specialization of labour and discovery of new goods and methods of production in a self-perpetuating process of economic growth. They are built on the basis of the idea that a given level of natural resource requires the use of labour, capital and the "effectiveness of labour" (technology) to spring-up a production process. At any given time the economy has some amounts of capital, labour and knowledge that are combined to produce a given level of output, implying that changes in input over time leads to changes in output correspondingly. Regardless of the levels of any factor of production, technological change is the only factor that can change per capita long-run growth rate of an economy. Hence, the "effectiveness of labour" (knowledge or technology) is the fundamental determinant of high level of sustainable economic growth.

There are four variables of production; output $(Y)$ and three inputs; capital $(K)$, labour $(L)$ and "knowledge" or the "effectiveness of labour" $(A)$ that enters a production function of constant returns to scale in the form:

$$Y_{(t)} = F(K_{(t)}, A_{(t)}L_{(t)}) \tag{1}$$

At any point in time $(t)$ the economy has some amounts of capital, labour and knowledge that are combined to produce a given level of output. Change in input over time leads to corresponding change in output. Based on the multiplicative effect of technology $(A)$ and constant returns to scale, it is possible to denote $k$ as capital per unit of effective labour and $y$ as output per unit of effective labour leading to intensive form of the production function as follows:

$$y_{(t)} = f(k_{(t)}) \tag{2}$$

The growth rate of labour is represented by $n$, that of knowledge (technological change) by $g$ while $\delta$ represents the rate of depreciation of capital. Moreover, output is partly consumed and partly saved at any given point in time, which give rise to the formation of change in capital stock as:

$$\dot{K}_{(t)} = sY_{(t)} - \delta K_{(t)} \tag{3}$$

In intensive form and using the definitions of $n$, $g$ and $\delta$, change in capital stock per unit of effective labour can be expressed as:

$$\dot{k}_{(t)} = sf\left(k_{(t)}\right) - \left(n + g + \delta\right)k_{(t)} \tag{4}$$

$\dot{k}_{(t)}$ is the rate of change of the capital stock per unit of effective labour

$sf(k_{(t)})$ is the actual investment per unit of effective labour, and

$(n+g+\delta)k$ is break-even investment

According to (4), the rate of change of the capital stock per unit of effective labour at any time is the difference between actual investment per unit of effective labour and the break-even investment. The break-even investment is necessary for two important reasons; to replete the depreciation of existing capital and to respond to growing quantity of labour to sustain or enhance its effectiveness. For the break-even investment to be adequate to match this requirement, it must be equal to the sum of depreciation rate and the rate of growth of the quantity of effective labour $(n+g+\delta)$.The capital stock per unit of effective labour will be rising whenever actual investment per unit of effective labour, $sf(k)$ exceeds the break-even investment, $(n+g+\delta)$ and vice versa, and when the two are equal, capital stock is constant.

The economy, regardless of its starting point, eventually converges to a balanced growth path, where all the variables grow at a constant rate and at this stage, the growth rate of output per worker, a key measure of economic growth, is determined only by the rate of technological change. Changes in all other variables, apart from technological change, will only lead to a shift in the level of the balanced growth path.

The key conclusions of the neoclassical growth analysis implies that, differences in capital per worker (K/L) and differences in the effectiveness of labour are the two main sources of variations in economic growth over time and across countries. However, only the changes in effectiveness of labour, which occurs through technological change, can generate permanent growth. Significant changes in saving have only moderate effects on the level of output per unit of effective labour on the balanced growth path, but not on the growth rate of the economy. Capital per worker influences output per worker, so a country that saves more of its output has more capital per worker and hence more output per worker but requiring the strong effect of technological change to stimulate high sustainable growth of the economy.

## 2.3 Endogenous growth theory

The main motivation for endogenous growth theory is to identify how technological change can occur from economic activities, rather than exogenous factors adduced by the neoclassical economic growth theory. Technological change evolves from the interplay of economic forces in a two-way interaction between technology and economic life. Technology is a by product of innovation, which is nurtured by rational economic behaviour; but technology also transforms economic life in turn. Ideas are the root of technology, which can be obtained from the production process as factors of production, especially labour, tend to learn and know more through engagement in production activities and seek to improve over time. This facilitates technological change through learning-by-doing. Incentives for high share of markets motivate firms to invest in Research and Development (R&D) to build on learning-by-doing advantages to bolster the momentum of technological change that leads to improvements in quality of products and emergence of new products.

Endogenous technological change emanates from three main sources; accumulation of physical and human capital, learning-by-doing and R&D. A firm that increases its physical capital learns simultaneously how to produce efficiently due to technical knowledge embodied in new capital goods. Each time a capital good is produced, the experience of producing it generates new insights to both the particular production sector and to the economy in general. It implies that investment and production makes use of ideas and also obtains additional ideas through the positive effect of production experience, thereby eliminating the tendency for diminishing returns of factors of production, making it possible for technological change to occur as intended or unintended by-product of investment. Hence, technological change is endogenous because it evolves from the operations of the economy.

Absence of diminishing returns to capital makes it possible for per capita growth to occur in the long-run driven by technological change that emerges from economic activities. There is one-to-one relationship between output and inputs due to constant returns to scale. One unit of either physical or human capital input leads to one unit of (additional) output. The non-diminishing marginal product of capital give rise to production function of the form:

$$Y_t = A_t K_t \tag{5}$$

$A$ is a positive constant reflecting the level of technology. In intensive (output per worker) form:

$$y_t = A k_t \tag{6}$$

$K$ is conceived broadly to encompass physical and human capital, knowledge and public infrastructure. The growth rate of investment per unit of effective labour is in the form:

$$\frac{\dot{k_t}}{k_t} = \frac{sf(k_t)}{k_t} - (n + g + \delta) \tag{7}$$

Due to learning-by-doing and other associated positive externalities, this prevents the marginal product of capital from diminishing, hence the production function of the form:

$$Y = F(K_i, A_i L_i) \tag{8}$$

Several firms engage in investment based on this production function. An increase in a firm's capital stock leads to a parallel increase in its stock of knowledge through learning-by-doing. Each firm's knowledge is assumed to be a public good, so other firms can gain access to it at zero cost. This implies that knowledge spills over onto the entire economy so each firm's discovery of new knowledge (technological change) is a reflection of the level of technology of the overall economy and is therefore proportional to the change in the aggregate capital stock. Learning-by-doing and knowledge spillovers make it possible to replace $A_i$ with $K$ and for the production function to be written in the form:

$$Y = F(K_i, K.L_i) \tag{9}$$

Each firm expands its capital stock, $K$, in the process of production, so $K$ rises accordingly and provides a spillover benefit that raises the productivity of all the firms, thereby generating endogenous growth. Each firm's increase in its capital stock adds to aggregate capital stock, and hence contributes to the productivity of all other firms in the economy.

Activities of government through its functions of public expenditure for the provision of infrastructure services, the protection of property rights as well as taxation policies have implications for technological change. Assuming there is no population growth and based on the activities of government, the aggregate production function will be in the form:

$$Y_t = AL_t^{1-\alpha}K_t^{\alpha}G_t^{1-\alpha} \tag{10}$$

This exhibits constant returns to scale in the private inputs, $L$ and $K$. If G (government expenditure) is fixed, there will be diminishing returns to capital accumulation, $K$, except if G rises along with $K$. This implies that public services are complementary with the private inputs in the sense that an increase in G raises the marginal products of labour and capital while the exponent of G $(1-\alpha)$ determines the extent to which G impacts on technological change to drive economic growth. For instance, if the exponent of G is less than $(1-\alpha)$, there will be diminishing returns to K and G and this will stultify technological change and by extension, endogenous growth.

Technological change is further enhanced when firms, driven by profitability, invest their resources in R&D leading to either quality improvement or variety expansion. Technology is regarded as a private product, so investors enjoy some level of preservation either because of the possibility of secrecy or acquisition of patent rights. Innovation leads to new products either in quality or variety, so innovators exploit some form of monopoly power. It is assumed that there are no bounds to new ideas, so there is no diminishing return in the creation of technology.

The final output and the R&D sectors as well as the labour market are assumed to be competitive, but the intermediate goods sector that provides inputs to the final goods sector is based on blueprints from R&D (knowledge). As R&D success leads to a new "state-of-the-art" version of the products through innovation, an existing product is replaced by an improved version of it or completely different version rendering it obsolete. Since the newly invented product will be available in the market, other researchers can examine its characteristics and learn knowledge embodied in it and use it for further research that could lead to further innovation of an improved version of it. This is a case of knowledge spill-over, which brings to the fore, the non-rivalry and non-excludability attribute of knowledge. This process is described as "Quality-Ladder" phenomenon or "Creative Destruction" (Schumpeter, 1975).

## 2.4 Inferences and deductions

The analytical building blocks of the two main economic growth theories (neoclassical and endogenous) implies that, a baseline technology is a key input that provide an initial condition for appropriate mix of factors of production. This lends credence to the fact that, it is the value-adding capabilities of the factors of production as a result of their effective use in production process that generates economic growth. Even though they both underline the essence of technological change as the driving force of economic growth, they differ on the sources and mechanisms through which technological change occur to impact on economic growth. The neoclassical theories subscribe to an exogenous (external) technological change effect while the endogenous proponents emphasize the emergence of technological change from active involvement in economic activities.

It is possible for technology transfer (exogenous technological change) to catapult economies with low level of technology to achieve high levels of sustainable growth (Bernard and Jones 1996; Dowrick and Rogers 2002). However, this will require adaptation of transferred technology into the stream of economic activities to provide a basis for "learning-by-doing" that diffuses into various sectors of the national economy to propel technological change. Some other economies can grow through the transfer of existing ideas as well as positive externalities of production processes. This reflects the proper application of ideas as a contingent part of the growth process, incorporated as a factor of production with a balanced need for using existing ideas and producing new ideas (Romer, 1992).

The significance of labour input in the production function means that there could be a positive relationship between the size of the population and economic growth. This has given rise to the idea of "scale effects" in economic growth analysis but mere size of population without developing and appropriately utilizing capabilities in production process does not provide significant advantage for technological change and economic transformation. This implies that, a foremost condition for optimal utilization of technological knowledge is development of robust human capital to be complemented by opportunities to unleash human capital in pushing the frontiers of technological change.

Thus, knowledge-in-use, not knowledge per se is critical for engendering technological change and the nature and dimension of knowledge spillover effects determine the robustness of economic growth. Therefore, effective number of researchers, rather than the population, is the critical determinant of production of ideas. In essence, high sustainable economic growth, which is the fountain of economic transformation, hinges on significant increase in productivity, which, in turn, depends on technological change that emanates from new ideas (designs) through R&D that springs from the labour force, which is a function of human capital that is drawn from population.

The productivity of competitive firms depends on their ability to innovate to adapt to technological change in order to gain from markets. Technological change festers on absorptive capacity (ability of capital investment or resource to yield appreciable level of return) of the overall economy. The absorptive capacity of the economy drives endogenous demand through the use of goods and services of a sector by other sectors of the economy. The essential relationship between effectiveness of labour and technological change requires intensive and efficient utilization of outputs of different sectors by other sectors of the economy. Thus, the intensity of sectoral interdependence generates high level of learning-by-doing and prompts the need for innovation that leads to R&D activities, which springs into technological change and economic transformation.

Apart from the neoclassical and endogenous growth theories, other perspectives of economic growth analysis converge on the critical relationship between technological change and economic transformation. For instance, the evolutionary growth theory asserts that economic activities evolves and springs into economic transformation through natural interdependence between changes in aggregate demand and technological change (Foley and Michl, 2011). Moreover, the process of transformation growth hinges on structural changes of the evolution of an economy that is driven by growth in effective demand that in turn, stimulates investments through adaptation to technological change to respond to market needs (Gualerzi, 2011). Inferences from Classical-Marxian evolutionary model points to the fact that technological change results from a random neutral innovation process that follows competitive market behaviours and motivation for profitability with labour productivity and wages evolving in concert (Levy and Dumeril, 2011). Heterodox Growth Theories (HGT) has illuminated certain dimensions, the key among which is the distinction between natural and actual economic growth rates implying that growth is exhaustible in the long-run, if all potential factors of growth are fully and efficiently utilized (Setterfield, 2009).

The impact of technological change on economic transformation requires a commensurate change in key factors of production, especially capital and labour, to enhance Total Factor Productivity (TFP) of the economy. It implies that transitional growth rates will differ among economies based on differences in the ratios of capital to effective labour. Economies with lower ratios of capital to effective labour relative to the steady state values will grow faster. If different economies have the same parameters for taste, technology and population growth rate, variation in growth rates will occur due to variations in distances from steady state and the rate of decrease of returns to capital, which in turn depends on technological change.

## 3. Dimensions of technological change and economic transformation

### 3.1 Invention-innovation-diffusion mechanism

Technological change occurs through a three chain relationships-- invention, innovation and diffusion. Invention is the creation of an item based on original ideas and knowledge more often described as "breakthrough" technology. Innovation is additional creativity that improves the features and usefulness of invented products. Diffusion refers to the spread of technological knowledge into various streams of economic activities that expands the space for further creativity to amplify the chain mechanism of invention, innovation and diffusion. Each aspect of this chain involves the appropriation of ideas through direct acquisition, "learning-by-doing" and R&D. Innovation is the pivot of this chain relationship in that innovation inspires invention and motivates diffusion, implying that both invention and diffusion posses some attributes of innovation.

An economy without the requisite technological wherewithal needs to evolve a system of innovation to engender technological change, which is synonymous with knowledge. System of innovation entails a network of institutions in the public and private sectors whose activities and interactions initiate, import, modify and diffuse new technologies (Freeman, 1987). As the pivot of the chain relationship of technological change (invention-innovation-diffusion), innovation constitutes the bedrock of system of innovation.

The essence of system of innovation is that even though natural endowments confer strategic advantages for certain activities that relates to specific aspects of technological knowledge, it is possible to create the requisite conditions for activities to flourish and propel technological change. For instance, geographical agglomeration (concentration of people and activities) is essential for stimulating effective diffusion and accumulation among local firms through which the process of technological change can be skewed to reflect the functional performance of firms, sectors, countries and regions based on the efficiency of the institutions that embody the innovation system (Patrucco, 2005). This underlines the crucial role of institutions in achieving technological change in firms, sectors, countries and regions. A functional system of innovation amplifies technological change due to penetration of knowledge into various economic activities.

Innovation is therefore the most critical factor in transforming sectors into dynamic systems through adaptation and interaction of factors of production based on coordinated national system of innovation. All agents within the innovation system are active partakers in the process of learning as economic activities continue over time. Learning takes place heuristically over a long period of time and possesses an incremental character. Technological change emerges as a by-product of active participation in the process of production, complemented by R&D activities. The system of innovation could target a key sector and after success spill-over to other sectors of the economy. To achieve this, the institutional processes, functions and policies need to play crucial roles of recognizing the essence of evolving cognitive technological capability to enhance the value-adding performance of factors of production.

Knowledge and by extension, technology, posses some degree of two significant attributes of public good; non-rivalry and non-excludability. A purely rival good is that for which its use by one agent precludes its use by another. In the same vein, a purely excludable good is that for which the possessor is capable of preventing others from using it either through legal means (property rights) or secrecy of inbuilt knowledge. Complete rivalry and excludability is not applicable to knowledge and technological change given that in the three chain processes (invention, innovation and diffusion), sharing with, and learning from other sources is imperative. Ideas that originate from thinking and reflections on the need to create something for specific use in real life is invention, which forms the foundation upon which other aspects of technological processes are built. Continuous use of invented technological product leads to innovation based on improvements in the form, use and adaptation of the initial product. The diffusion mechanism takes cognizance of knowledge or technology that is useful to the activities and sectors within the economy.

That is, knowledge spill-over effect is made possible by the degree of non-rivalry and non-excludability of knowledge that is imbued in technology. Any newly invented product will be available in the market, researchers can examine its characteristics and learn knowledge embodied in it so as to replicate or produce improved version of it through R&D. The non-rivalry character of knowledge and the possibility of spillover benefits to rival firms implies that the gains of R&D (profitability) is not limited to one firm alone making it possible for the social benefits to outweigh the private ones. Empirical study of innovating German manufacturing reveals that incoming spillovers have a positive effect on profitability on top of a firm's own R&D investment (Czarnitzki and Kraft, 2012). Improvements in quality of innovation leads to continuous replacement of existing products with a new version and the old ones become obsolete continuously over an infinite horizon. A key sector with innovation-driven technological change interacts with other sectors of the economy to expand opportunities for innovation and technological change through adaptation that is anchored on interconnectedness of sectors of the economy.

Technological inter-connections among various sectors of the economy evolve from structural and spatial interdependence of the production processes of the sectors. Through rational response to inducements and incentives, capabilities of factors of production are enhanced and transmitted into technological relationships. The cumulative effect is that as the sector from which technological "breakthrough" occurs increase production to take advantage of the efficiency of new technology, it increases the production of other sectors through demand for raw materials. In response to the demand from the resurgent sector, other sectors will seek to improve on their delivery efficiency and in the process adapt to existing technology or improved version of it. Economic activities will expand across sectors of the economy through this self-reinforcing process of inter-sectoral linkages. This provide opportunities for economies of scale that leads to lower per unit cost of production, thereby translating into market advantages that propel industrialization as manufacturing activities expand to take advantage of global markets.

## 3.2 Industrialization, manufacturing and globalization

Economic development experiences of both advanced and emerging countries illuminate the fact that technological change is the most critical factor in the transformation of low-level economies into high-level economic activities to drive sustainable high economic growth path. Technological change enhances human capabilities that lead to quality improvements and efficiency in terms of producing more without additional resources. Corollary, increasing specialization of labour leads to discovery of new goods and methods of production. This enhances Total Factor Productivity (TFP), which propels the "effectiveness of labour" that brings about high sustainable economic growth path. It has been established that differences in developmental levels of countries are largely due to differences in the efficiency of production as measured by relative levels of TFP, which is a reflection of technology gap (Hulten and Isaksson, 2007).

The need for technological change to drive economic transformation has become even more intensified by an increasingly interdependent global economic dispensation that tends to undermine and marginalise indolent economies. Economic growth disparities among countries of the world are largely attributable to degree of technological change that has, through improvements in the effectiveness and efficiency of harnessing economic resources, impacted on economic activities. This has given rise to a "four-speed world" categorisation of the global economic landscape relative to economic growth transformation achievements into four group of countries (Wolfensohn, 2007) as follows:

**Affluent:** those that have maintained global economic dominance for 50 years, constituting about 20 percent of world's population yet accounting for about 70-80 percent of global income. The United Sates of America, Canada, Japan and Germany are some of the countries in this group.

**Converging:** those that are poor and middle income economies but achieve high and sustainable economic growth (economic transformation) to emerge as key global economic players. China and India are prominent in this group.

**Struggling:** those with unsteady pattern of economic performance (irregular growth achievements at times strong and at times low), which implies a lack of coherent economic structure that drives economic performance. Their influence on global economic system is relatively weak.

**Poor:** those countries where income is stagnating or falling, mostly in Sub-Saharan Africa and where most of the "Bottom Billion" (Collier, 2007) lives. In the context of global economic schemes, they are very weak and significantly vulnerable to the adverse effects of globalisation.

The most significant underpinning factor for the differences in economic performances in the context of the four-speed world is the capacity to generate and absorb new technologies, which reflects in TFP of countries. The growth of TFPs of different group of countries in the four speed categorization over the period 2000-2007 (Table 1) indicates that the converging group of countries has the highest TFP growth of 2.8 percent, the affluent with 1.1 percent while the struggling and poor groups achieved 0.5 and 0.6 percents respectively (OECD, 2010). This indicates a growing technological divide that underpins the transformation of global industrial and economic landscape that is springing new global economic power houses. This is even more obvious from the TFP growth rates of China and India, the most remarkable success of economic transformation in recent years, at 4.4 and 2.1 percents respectively. As Table 1 further illustrates, there is a strong relationship between TFP growth and output growth as well as strong correlation between TFP growth and other key factors of economic growth; physical capital growth and human capital growth.

| | Output growth | TFP growth | Physical capital growth | Human capital growth |
|---|---|---|---|---|
| Affluent | 3.5 | 1.1 | 3.6 | 1.0 |
| Converging | 6.5 | 2.8 | 4.5 | 2.0 |
| Struggling | 3.0 | 0.5 | 2.7 | 2.6 |
| Poor | 3.0 | 0.6 | 2.8 | 2.7 |
| Brazil | 3.4 | 1.4 | 1.6 | 2.4 |
| China | 9.3 | 4.4 | 9.6 | 0.9 |
| India | 7.0 | 2.1 | 8.1 | 2.2 |
| South Africa | 4.2 | 1.8 | 3.6 | 1.4 |
| OECD Average | 2.5 | 0.9 | 1.0 | 1.2 |

Source: OECD (2010), *"The Growing Technological Divide in a Four-Speed World"* in Perspectives on Global Development 2010: Shifting Wealth"

Table 1. Average Annual Growth of Key Economic Growth Factors, 2000-2007 (%)

Global development experiences have shown that, the most effective route to economic transformation is industrialization, the core of which is robust manufacturing activities. Historical evidence indicates that it is rare for any country to achieve high-sustained growth without industrializing as virtually all advanced economies experienced industrial revolution in their march towards development as stressed by the United Nation Industrial Development Organization (UNIDO) thus:

"Although the essence of industrialization is not new, recent changes in the global economy have substantially altered the opportunities for industrialization and recent academic research has, in turn, substantially changed our understanding of the process of industrialization and illuminates the significance of manufacturing. The past several decades have witnessed a major restructuring of the global economy, one in which more and more industrial output and employment is now located in emerging developing countries, while the developed countries have become ever more service oriented economies. Globalization, through increased trade and investment flows is driving this restructuring, along with technological and associated organizational change" (UNIDO, Industrial Development Report, 2009).

Manufacturing value-added (MVA) plays multiple roles in industrialization and economic transformation. It enhances productivity, increases absorptive capacity and provides the basis for "learning-by-doing" that springs into technological change. Cross-country economic growth empirics points to the fact that, structural change that leads to shift in capital and labour from low productivity to high productivity sectors by propelling the TFP, is the driving force of economic transformation[2]. Evidence of strong positive relationship between MVA and economic transformation abound. The IDR 2009 illustrates the crucial role of MVA in the stupendous economic transformation of emerging industrial countries that has transformed the global industrial and economic landscape. Based on long-term growth performance[3] and initial level of income in the base year between 1975 and 2005, the report classified countries into five different groups:

- High-income countries, mostly OECD member countries
- Fast-growing middle-income countries
- Slow-growing middle-income countries
- Fast-growing low-income countries
- Slow-growing low-income countries

Comparing the rate of MVA growth per worker in 1975-2000, MVA per capita in 1997 and 2005, and growth rates of countries, it revealed that MVA growth rates was about twice higher while MVA per capita was three times higher in fast-growing countries than in slow-growing ones. Thus, productivity gains in manufacturing accounts for the large differences between fast-growing and slow-growing countries. As further evidence of the significance of manufacturing, the emergence of East Asian industrial countries into high-growth economies has been due to large contributions of their manufacturing relative to other sectors.

The changing share of manufacturing in GDP of regions indicate that, there is strong causality between MVA and high growth performances of the Asian tigers that are driving global economic growth pattern in recent years. In East Asia, the relative share of manufacturing to GDP increased from 25 percent in 1965 to about 35 percent in the 1980s and remained at above 30 percent throughout the 1990s. Conversely, the manufacturing share of GDP in Latin America, which was at the same level with East Asia in 1965, remained stagnant. Furthermore, as a reflection of the critical role of MVA in the changing global economic landscape, emerging global economic powers achieved significant growth in MVA while that of developed countries has slowed. Due to the significant rise of emerging economies, average growth rate of MVA of developing countries have increased significantly above that of developed countries in recent years. For instance, average annual growth rate of MVA of developing countries between 2001 and 2005 was 6.2 percent against world average of 2.7 percent and developed countries average of 1.4 for the same period. Developing countries improved to 7.1 percent between 2006 and 2010 while developed

---

[2] While an economy's aggregate TFP is a weighted sum of each sector TFP levels, TFP growth of the entire economy reflects also, the changes in the structural composition of the economy. Lipsey and Carlaw (2004) used simple mathematical calculations to show that aggregate TFP changes as movement of labour between formal sectors occur.
[3] Growth performance is measured by "growth experience", which is defined in terms of GDP per capita growth above the median sample. Countries with "growth experience" more than half of their annual observations are classified as "fast growers".

countries deteriorated to 0.2 percent for the same period with the world average at 2.4 percent. China and India, the foremost in current global economic transformation, achieved average MVA growth rate of 4.8 and 8.6 percent for the period 2001-2005 and 4.9 percent and 6.2 percent for the period 2006-2010, respectively.

In a nutshell, the existence of industrial production on one hand, and demand for the products of the industries on the other hand, creates opportunities for market expansion, competition and specialization. Through a favourable "forward linkage" effects, an endogenous self-perpetuating process of growth emerges and feeds on it almost automatically. Through internal and external economies of scale, the process of industrial production evolves into higher and more sophisticated levels of production, giving rise to further specialization, new products and quality improvements, leading to technological change that spurs economic transformation. Adaptation to a growing market, widened by international trade, stimulates industrial production and provides additional impetus for technological change and economic transformation. Globalization facilitates both technological change and economic transformation as it creates opportunities for market expansion, competition and specialization, which are essential for industrialization. Extension of markets in integrating economies across national borders has contributed significantly in global technological change and economic transformation, especially through the catalytic effect of information and communications technology (Atkinson, 2009). The process of structural transformation increases the relative contribution of manufacturing activities with strong interdependence among domestic sectors and regional economies. This provides the basis for expansionary effect of inter-industry linkages and creates opportunities that enhance prosperity and standard of living.

### 3.3 Prosperity and standard of living effect

High sustainable economic growth is the *sine qua non* for economic transformation but prosperity relates to the share of the proceeds of economic growth benefits that reaches most of the people. Since benefits come in form of rewards for work, effective participation of most, if not all, of the people, in the productive activities of various sectors of the economy in the process of economic transformation is essential. Therefore, economic growth needs to be inclusive by providing equal opportunities to all members of the society to participate and contribute to the growth process regardless of their circumstances (Ali and Zhuang, 2007) to inspire the process of technological change and economic transformation. This will require that all productive sectors of the economy are active enough to absorb factors of production to optimal level through multi-sectoral input-output interdependence of productive activities.

The effect of technological change has been very tremendous not only on living standards but also in life characteristics. For instance, output per worker in the United States increased 10 times more than 100 years ago (Maddison, 1982). There has been a 7-fold increase in market TFP in the USA between 1800 and 1990 along with increase in real wages by a factor of 9 over the period 1890 to 1990, which led to a decline in fertility from 7 kids per woman in 1800 to 2 in 1990 (Greenwood and Seshadri, 2004). China's GDP has almost doubled between 1997 and 2004 with the annual change in GDP approaching 8.3 percent per year leading to increase in per capita income by 83 percent in urban areas and 41 percent in rural areas (Bromley and Yao, 2006). This fundamental changes in life patterns has been largely due to the fact that technological change has significantly improved the skills of workers, propelled industrialization and driven economic transformation.

In any modern society, the people are involved in four basic activities that are intertwined with their livelihood. These are:

- *Production* of goods and services by industry and commodity sectors at different stages of activity chain (primary, manufacturing and service activities).
- *Consumption* (purchases) of goods and services by industries, individuals and various government agencies which provide markets for the goods and services produced to create room for more and continuous production.
- *Trade* which involves selling goods and services produced by the society and buying of goods and services produced from elsewhere.
- *Accumulation* (generating surpluses) through savings and capital transactions such as fixed investment expenditure and stock change made possible by the surpluses generated after the production and selling of goods and services.

The intensity of involvement of the people and the level of technology applied in the process of production determine the level of value-added and by extension the level of income to be earned from the proceeds of trade. The higher the value-added content the higher the returns on investment and the lower the value-added content the lower the returns on investment. Higher level of returns could lead to surpluses (extra income after purchases of needed goods and services) which could be ploughed back into the investment stream (accumulation). This leads to a regeneration process that is self-perpetuating and thus a path towards self-reliant and sustainable economic prosperity that spills over to other aspects of life. As the process of production becomes more sophisticated and global markets lead to expansion of economic activities, the nature of labour participation in production changes due to the effects of technological change. For instance, empirical evidence suggests a link between declining share of labour in value added manufacturing and increasing productivity that is driven by expansion in international trade (Böckerman and Maliranta, 2012). This leads to higher productivity with declining labour involvement with much higher earnings relative to input-output ratio thereby enhancing prosperity.

Thus, economic prosperity is attainable through industrial production activities that cater for the essential needs of the people and provide opportunities for the people to work and earn income. The higher the value-added content through the application and adaptation to technological change, the higher the returns on investment and the lower the value-added content the lower the returns on investment. The higher the levels and intensity of the involvement of the people of a society, the more they are able to provide for their needs, trade favourably and generate surpluses and hence the more economic prosperity and higher standard of living the society can attain.

## 4. Fostering technological change and economic transformation

### 4.1 Essential conditions

Previous sections have established that technological change and economic transformation are mutually reinforcing aspects of development process that enhances standard of living. Effective utilisation of resources is fundamental to the attainment of technological change and economic transformation. It requires strategies that create incentives for investments that use natural resources as intermediate goods and transform them into finished goods by a manufacturing production process that enhances the value-adding capabilities of factors of production. Institutional efficiency and effective macroeconomic management are essential in creating solid infrastructures that forms the basis for fostering technological change and economic transformation.

Technology is related to production of all aspects of goods and services and has four components. The first component is human capital involving the training of people to equip them with skills. The second component is technical requiring the provision of necessary equipment and new materials. The third is institutional, which is about regulatory and policy framework and the tools of implementation. The fourth component is the informational aspect, which is about accessing available developments and progresses in global technological application. Each of these components is crucial for ensuring continuous improvement in the methods of production, discovery of new resources and thus creating the necessary conditions for efficient utilization of resources to foster technological change and economic transformation.

A system of innovation strategy is essential for ensuring effective domestic participation in value-adding activities to generate synergy for technological change and economic transformation. The role of the public sector is crucial in providing the requisite platform for generating ideas (knowledge or technology) through learning-by-doing and R&D activities, as well as co-ordinated linkages among sectors of the economy. This creates incentives for effective private sector investments that expand economic activities and opening-up opportunities for knowledge spillovers, learning-by-doing and R&D to engender technological change and economic transformation.

Technological change and economic transformation depends largely on the creation of ideas that are derived from human capital, which draws from population of a country. However, ideas do not automatically emanate from population as certain conditions are required for knowledge creation to occur. Hence, there is no direct correlation between population growth and technological change. In real world, there are examples of countries with both large and low population that recorded significant success of achieving technological change and economic transformation. The effective number of researchers, rather than the population, along with thriving competitive enterprise are the crucial driving forces of technological change and economic transformation. Population growth could be useful in providing a pool of human resources that can be effectively transformed through training and productive engagement to create ideas and steer the process of technological change.

Even though, the fundamentals of fostering technological change and economic transformation are familiar to a large extent, many developing countries, especially Least Developed Countries (LDCs) have not demonstrated significant achievement in fostering technological change and economic transformation. This is largely due to entrenched public sector inefficiency in management, coupled with weak production structures, which combine to constitute stumbling block for technological change and economic transformation. Various agents play essential roles in fostering technological change and economic transformation but the two most critical agents are the public and private sectors.

## 4.2 The role of the public sector

Efficient functioning of government in discharging its responsibilities to create enabling condition for investments and thriving industrial production create opportunities for expansion of economic activities, which enhances the efficiency and effectiveness with which economic resources are utilized. Availability of efficient public services provides incentives for industrial production that leads to expansion of economic activities through interconnectedness of economic activities of various sectors of the economy. Large involvements of people in chain economic activities create opportunities for enhancing their capabilities to motivate innovation instincts to drive technological change.

This conforms to the fundamental development principle that the economic and social progress of any country depend largely on government's ability to generate sufficient revenues to finance an expanding programme of essential, non-revenue yielding public services (Todaro, 1994). Human capital formation, which is the bedrock of economic transformation and technological change, needs to be provided or strongly supported by governments through non-profit making principles. Production activities by all sectors of the economy is possible only if basic infrastructures and the rule of law that guarantees property rights (patents and copy right laws) are in existence. Economic transformation occurs in the course of development as public sector activities evolve through a system of revenues that accrue to the government and expenditures based on the varying and changing needs of the economy.

Although essential services (infrastructure, rule of law and human capital formation) are needed by all levels (household, firms and government), their non-excludability character means that firms with competitive profit maximizing objectives would not like to finance their provisions. Furthermore, there is the need for effective coordination to strengthen the significant relationship between consumption and production that is anchored on input-output mechanism that accentuates industrial production. The functions of providing essential services (public goods), which includes critical coordination of economic and social activities to align with aspiration of the society are functions that can only be undertaken by government based on its non profit and welfare provision disposition.

Thus, effectiveness of government's coordination and essential services provisions is a crucial component of the building blocks of economic transformation and technological change. For instance, sound educational and health service delivery will lead to the emergence of skilled and healthy workforce, a prerequisite for "effectiveness of labour", which in turn, is a key requirement for industrial production that leads to economic transformation and technological change. Beside, natural resource sectors which are the basic seeds from which economic activities germinate requires legal and institutional framework of operations and this can only be provided by government institutions.

There is complementarily between public services and private sector activities in the process of economic transformation and technological change. A robust and efficient system of government expenditures leads to high marginal products of labour and capital to individual firms. This is because of the quality of capital formation that occurs due to good educational and health provision functions of government. Government purchases a portion of private sector outputs with which it uses to provide free public services that is non rival and non excludable. Firms benefit from this as effective source of demand for their goods and services to be able to meet the wage requirements of the highly skilled labour and as well enhance their profit levels. This reflects in the high quality and large quantity of the aggregate output of the economy. In this context, government is a key factor in facilitating economic transformation and technological change in addition to enhancing the prosperity of the people through social optimal growth of the economy.

Adequate provision of basic needs of the society through the public sector machinery motivates private investments and enhances the productivity of factors of production to stimulate high sustainable economic growth. Thus, effective functioning of government creates the foundation upon which robust private sector activities spring to build upon existing ideas associated with economic activities through "learning-by-doing" and R&D activities. This provides opportunities for the emergence and effective contribution of the private sector.

## 4.3 Private sector

Robust private sector activities are essential for technological change and economic transformation. It is widely perceived that private sector investments are more effective due to their higher efficiency in utilizing resources. Private sector investments tend to use ideas much more and gain additional ideas through the positive effect of production experience. Knowledge creation is an unintended by-product of investment as such as firms increase the combination of their physical and human capital; they automatically improve their efficiency in production beyond the equivalent levels of the increase in capital. Private sector firms therefore tends to apply technical knowledge in their activities more and therefore are more likely to gain insights from capital goods that enhances production activities in their particular sectors as well as the overall economy through spill over effects.

In their drive to gain large control of markets, firms invest in R&D to improve quality and expand variety of their products. New discoveries of technology by firms through the combination of learning-by-doing and R&D are initially regarded as a private product and they enjoy some degree of preservation that gives them a measure of monopoly power over the discovery with the support of patent rights, in addition to secrecy of methods and codes of the technology. However, over time, competitive firms gain access to the new products from the market and analyse the new knowledge in them and make efforts to replicate or apply the ideas to create their own products with equivalent or even higher level of sophistication.

Increase in a firm's capital stock leads to a parallel increase in its stock of knowledge through learning-by-doing. Each firm's knowledge is assumed to be a public good, so other firms can gain access to it at zero cost. This implies that knowledge spills over onto the entire economy so each firm's discovery of new knowledge (change in technology) impacts on the overall economy. Therefore, a firm's level of technology and by extension, the changes in its technology, is proportional to that of the overall economy.

There are two dimensions of spillover effects of technological change on private firms. The first is sustaining technologies, those that help organizations to make marginal improvements in what they do and requires only gradual change to modify existing systems and products. The second is termed disruptive technologies, those that involve fundamental and at times unexpected technological breakthrough that requires corporations to radically rethink their very existence. There is the tendency for large companies to be comfortable with sustaining technologies due to their preoccupation with maintaining markets under their control and aversion to risks of uncertainties associated with disruptive technologies. Apart from the challenge of successful application of the newly "unknown" disruptive technologies, getting consumers to accept the transformed product could threaten market control (Christensen, 1997 and 2003).

Firms that pioneer disruptive technologies do not usually achieve straightforward success in transforming their products and getting the most market attractions and therefore could experience deteriorating performances. However, nurturing the process of applications leads to improvements that make it possible for disruptive technologies to bring substantial benefits to firms in terms of market shares based on positive sentiments of new designs and the perception that it is associated with quality improvements. Products that emerge from disruptive technologies are therefore of limited interests because they don't provide "quick wins" for firms but for those that endure the gradual process of mutation of the technologies into new products, they eventually completely overtake existing products and markets thereby bringing substantial benefits to endured firms.

The comfort of short-term gains from sustaining technologies and averting the risks of uncertainties associated with disruptive technologies on one hand, and the potential substantial gains of the success of nurturing the applications of disruptive technologies on the other, presents a dilemma to firms. To resolve this dilemma, firms will need to acquire separate "spin-off organizations" that are separate from their mainstream operations to nurture the applications of disruptive technologies to eventually reap associated benefits. This strategy provides a basis for accommodating failures from applications of disruptive technologies. (Christensen, 1997 and 2003)

## 4.4 Key conclusions and insights

Fostering technological change and economic transformation could address a pertinent intergenerational economic management dilemma; a trade-off between present and future consumption and by extension, welfare. The declining effect of the use of economic resources especially exhaustible natural resources, over time implies that output and consumption will also decline over time. Sustaining a level of output and consumption over a long period to establish intergenerational equity therefore becomes a challenge.

The combined effect of technological change and economic transformation ensures that through inventions, innovations and diffusions, quality of available goods and services are enhanced and new goods and services are created through more efficient production processes. In the context of sustainability for intergenerational prosperity, technological change could ensure non-declining consumption (utility); maintaining (constant) production opportunities over time; non declining natural capital stock; maintaining a steady yield of resource services; stability and resilience of the ecosystem through time and the development of capacity for consensus building.

A key strategy in fostering technological change and economic transformation is significant investments in human capital development through various aspects of education, training for skills acquisition and provision of social support services. This needs to be in tandem with facilitation of large-scale investments that create expansionary effects on the economy complimentarily with a robust household sector that provides effective market and supply labour services. Effective functioning of the public sector in providing essential needs propels the private sector to engage in profitable investments that in turn, motivates the drive for technological change through direct acquisition, learning-by-doing and R&D activities.

Efficacy of regulatory framework, the structure and operations of critical sectors of the economy along with sound macroeconomic management of the economy is essential. Appropriate measures for acquiring technological capabilities should be anchored on domestic value-adding activities based on formidable linkages between a strategic sector with other sectors of the economy driven by innovative transformation of sectors into dynamic systems through adaptation and interaction of factors of production based on co-ordinated national system of innovation.

## 5. Acknowledgments

The author is grateful to the comments and suggestions of the anonymous reviewer, which has helped improved the contents of the chapter. The contents of the chapter are the responsibility of the author and not his affiliated organization, the Islamic Development Bank.

## 6. References

Aghion, P. and Howitt, P. (1998) Endogenous Growth Theory. MIT Press, London, England.

Ali, I. and Zhuang, J. (2007): *"Inclusive Growth Toward A Prosperous Asia: Policy Implications"* Asian Development Bank, ERD Working Paper No. 97

Atkinson, R (2009): "Globalisation, New Technology and Economic Transformation" in Cramme, O. and Diamond, P (eds), Social Justice in the Global Age, Polity Press, Cmbridge CB2 1UR, UK

Barro, R. and Sala-i-Martin, X. (1995), Economic Growth, McGraw Hill, New York.

Bernard, A.B. and Jones, C.I. (1996); *"Productivity across industries and countries: Time series theory and evidence"*. Review of Economics and Statistics, vol. 78, pp. 135-146.Bernard, A. and Jones

Böckerman, P. and Maliranta, M. (2012): Globalization, creative destruction, and labour share change: evidence on the determinants and mechanisms from longitudinal plant-level data , Oxford Economic Papers 64(2): 259-280

Bosworth, B. et al (1972) The Recent Productivity Slowdown: Comments and Discussions, *Brookings Papers on Economic Activity*, 3, 537-545.

Bower, J. and Christensen, C. (1995), "Disruptive Technologies: Catching the Wave", *Harvard Business Review*, January–February 1995

Bromley, D. W and Yang Yao, Y. I. (2006) "Understanding China's Economic Transformation: *Are there lessons here for the developing world?*", World Economics, Vol. 7, No. 2, April–June 2006

Chenery, H. B., Shishido, S. and Watanabe, T. (1962) The Pattern of Japanese Growth, 1914-1954 *Econometrica*, Vol. 30, No 1, January 1962, pp98-139.

Chenery, H. B., Robinson, S. and Syrquin, M. (1986) "Industrialization and Growth: A Comparative Study" The World Bank, Oxford University Press

Christensen, C. (1997), The Innovator's Dilemma: When New Technologies Cause Great Firms to Fail, *Harvard Business School Press,*

Christensen, C. and Raynor, M. (2003), "The Innovator's Solution", Harvard Business School Press,

Collier, P. (2007), "The Bottom Billion: Why the Poorest Countries are Failing and What Can Be Done About It", Oxford University Press

Currie, L. (1974) The 'Leading Sector' Model of Growth in Developing Countries, *Journal of Economic Studies,* Vol.1, No. 1, May, 1997.

Czarnitzki, D. and Kraft, K. (2012) Spillovers of innovation activities and their profitability, Oxford Economic Papers 64(2): 302-322

Dasgupta, P. and Stiglitz (1981), Resource Depletion under Technological Uncertainty. *Econometrica,* 49(1), 85-104.

Dowrick, S. and Rogers M. (2002), Classical and Technological Convergence: Beyond the Solow-Swan Growth Model. *Oxford Economic Papers,* 54, 369-385.

Dumenil, G. and Levy, D. (2011), "The Classical-Marxian Evolutionary Model of Technical Change: Application to Historical Tendencies" In: Setterfield, M. (ed) Handbook of Alternative Theories of Economic Growth, *Edward Elgar Publishers*

Foley, D. K. and Michl, T. (2011), "The Classical Theory of Growth and Distribution" In: Setterfield, M (ed) Handbook of Alternative Theories of Economic Growth, *Edward Elgar Publishers*

Frankel, M. (1962), The Production Function in Allocation and Growth: A Synthesis, *American Economic Review* 52 (December), 955-1022.

Freeman, C. (1987) "Technology Policy and Economic Performance: Lessons from Japan, Frances Printer, London

Greenwood, J. and Seshadri, A. (2004) Technological Progress and Economic Transformation, Working Paper 107665, National Bureau of Economic Research

Gualerzi, D. (2011), "The Paths of Transformational Growth" In: Setterfield, M. (ed) Handbook of Alternative Theories of Economic Growth, *Edward Elgar Publishers*

Harrington, D. R., Khanna, M. and Zilberman, D. (2005) Conservation Capital and Sustainable Economic Growth, *Oxford Economic Papers,* 57, 2, 336-359.

Hulten, C. and Isaksson (2007): "Why Development Levels Differ: The sources of Differential Economic Growth in a Panel of High and Low Income Countries", NBER Working Paper, No. 13469

Ishikawa, J., Sugita, Y. and Zhao, L. (2005), "Corporate Control, Foreign Ownership Regulation and Technology Transfer". Presented at FAI seminar, Department of Economics, University of Strathclyde, Glasgow, United Kingdom.

Jones, C. I. (1995), R&D-Based Models of Economic Growth, *Journal of Political Economy,* 103, 759-84

Jones, C. I. (1999) Growth With or Without Scale Effects? *American Economic Review Papers and Proceedings,* 139-144.

Jones C.I (2002), Introduction to Economic Growth, W. W. Norton, USA.

Koopmans, T. C. (1973) Some Observations on 'Optimal' Economic Growth and Exhaustible Resources In: Box, H. C., Linnemann, H. and Wolff, de P., *Economic Structure and Development,* Elsevier, New York, U. S. A., 239-255.

Koopmans, T. C. (1974) A Proof for the Case Where Discounting Advances the Doomsday, Review *of Economic studies,* 41, *Issue on the symposium of Exhaustible Resources.*

Koopmans, T. C. (1978) The Transition from Exhaustible to Renewable and Inexhaustible Resources, *Cowles Foundation Discussion Paper*, No. 486, Yale University.

Kubo, Y. and Robinson, S. (1984) Sources of Industrial Growth and Structural Change: A comparative Analysis of Eight Economies, *Proceedings of the Seventh international Conference on Input-Output Techniques,* United Nations, New York, 1984.

Kuznets, S. (1971) Modern Economic Growth: Findings and Reflections, *Nobel Prize Lecture,* December 11, 1971

Li, C. W (2001) Growth and Scale Effects: The Role of Knowledge Spillovers, *Economic Letters.*

Lucas, R. Jr (1990) Why Doesn't Capital Flow from Rich to Poor Countries? *American Economic Review,* 80, 92-96.

Maddison, A. (1982): Phases of Capitalist Development, Oxford University Press, Oxford

McCallum, B.T. (1996), "Neoclassical vs Endogenous Growth Analysis: An Overview", *Federal Reserve Bank of Richmond Economic Quarterly, Fall 1996. Also NBER working paper no. 5844.*

Organization for Economic Co-operation and Development (OECD, 2010); "The Growing Technological Divide in a Four-Speed World" in *Perspectives on Global Development 2010: Shifting Wealth,* OECD Report.

Patrucco, P. P. (2005) The Emergence of Technology Systems: Knowledge Production and distribution in the case of the Emilian Plastics District, *Cambridge Journal of Economics,* 29, 1, 37-56

Ramsey, F. P (1928) "A Mathematical Theory of Saving", Economic Journal, 38, 543-549.

Romer, M. P. (1992) "Two Strategies for Economic Development: Using Ideas and Producing Ideas" Proceedings of the Annual World Bank Conference on Development Economics.

Romer, D (1996) Advanced Macroeconomics, McGraw Hill.

Romer, M. P. (2007). Endogenous Technological Change, The Journal of Political Economy , Vol. 98, No.5 (March 200) pp. 71-102

Sala-i-Martin, X, Doppelhofer, G. and Miller, R.I (2004) Determinants of Long-Term Growth: A Bayesian Averaging of Classical Estimates (BACE) approach. *American Economic Review,* Vol. 94, No 4, September, 2004.

Sandilands, R. J. (2000). Perspectives on Allyn Young in Theories of Endogenous Growth. *Journal of History of Economic Thought,* Vol.22, No.3.

Schumpeter, A. J. (1975): "Capitalism, Socialism and Democracy", Harper and Row

Setterfield, M (2009) "Neoclassical Growth Theory and Heterodox Growth Theory: Opportunities For and Obstacles To Greater Engagement", Working Paper 09-01, Department of Economics, Trinity College, Hartford Connecticut, USA

Sinha, D. (1999) The Relevance of the New Growth Theory to Developing Countries In: Dahiya, S. B. (ed) *The Current State of Economic Science,* Vol.5, pp 2457-2466.

Solow, R. M. (1956), A Contribution to the Theory of Economic Growth. *The Quarterly Journal of Economics,* 70 (1), 65-94.

Solow, R. M (1994), Perspectives on Growth Theory, *The Journal of Economic Perspectives,* 8(1), 45-54.

Swam, T. W. (1956). Economic Growth and Capital Accumulation. *Economic Record,* 32, 334-361.

Todaro, P. M. (1994) Economic Development. Longman, New York

United Nations Industrial Development Organization (UNIDO, 2009), "Breaking In and Moving Up: New Industrial Challenges for the Bottom Billion and Middle Income Countries": *Industrial Development Report 2009*

United Nations Industrial Development Organization (UNIDO, 2011), "Industrial Energy Efficiency for Sustainable Wealth Creation: Capturing Environmental, Economic and Social Dividends": *Industrial Development Report 2011.*

Wolfensohn, J. (2007), "The Four Circles of a Changing World", International Herald Tribune, June, quoted in OECD (2010) "Shifting Wealth and the New Geography of Growth" in Perspectives on Global Development 2010: Shifting Wealth, OECD Report.

Wyckoff, A., Sakurai, N. and Leedman, C. (1992). Structural Change and Industrial Performance: A seven country Growth decomposition Study. OECD, Paris Cedex 16, France

Young, A. (1991). Learning by Doing and the Dynamic Effects of International Trade. *Quarterly Journal of Economics,* 106 (2), 369-406.

# 2

# Intellectual Property Rights and Endogenous Economic Growth – Uncovering the Main Gaps in the Research Agenda

Mónica L. Azevedo, Sandra T. Silva and Óscar Afonso

*CEF.UP, Faculty of Economics, University of Porto, Porto*
*Portugal*

## 1. Introduction

Intellectual Property Rights (IPRs) are "the rights to use and sell knowledge and inventions" (Greenhalgh and Rogers, 2007: 541), with the aim of guaranteeing adequate returns for innovators and creators. There are different types of intellectual property protection (Granstrand, 2005): old types such as patents, trade secrets, copyrights, trademarks and design rights, and new forms such as breeding rights and database rights. Nonetheless, patents are commonly considered as the most important and representative IPR (*e.g.*, Besen and Raskind, 1991).

IPR have a long legal and economic history, since the idea of intellectual property was already present in ancient cultures such as Babylonia, Egypt, Greece and the Roman Empire. Mokyr (2009) discusses the relevance of the late 19th century, when political events created a system which supported an executive that was sufficiently well-organised to create a "rule of law" and respect private property rights. This argument emerges, in part, in the context of an Industrial Revolution marked by important technological improvements, whereby IPR began gradually to be accorded more respect.

Despite this long history, only recently has IPR come to play a central role in debates concerning economic policy, being a stimulus for innovation through monopoly power (Menell, 1999). This change, related to the pro-patent era, only emerged in the 20th century – first in the USA and then globally in the world. Beneath this profound transformation lay a "deeper, more broad-based and much slower flow of events towards a more information- (knowledge) intensive and innovation-based economy" (Granstrand, 2005: 266). Therefore, in this period, knowledge and information assumed an important role in economics, which implied important changes in policy-making both in developed and developing countries.

The relationship between IPR, technological change and economic growth is ambiguous (*e.g.*, Horii and Iwaisako, 2007; Harayuma, 2009; Panagopoulos, 2009). Although knowledge and innovation are crucial for economic growth (*e.g.*, Romer, 1990; Aghion and Howitt, 1992; Hall and Rosenberg, 2010), if they are (completely) free there will be no incentive to invest in new knowledge and inventions (*e.g.*, Arrow, 1962; Romer, 1990). Thus, the potential need to protect both knowledge and inventions emerges, and the discussion of the importance of

IPR for this protection function gains relevance. In forming the decision whether to protect or not, a typical trade-off emerges: if we protect, only the owner of the knowledge design will use it (for some years she/he will have the monopoly power) and so the impact on economic growth will be smaller; in cases where no protection exists (which would allow innovators to be rewarded), knowledge will be easily diffused and all adopters will benefit from associated profits without having supported the corresponding costs; in the latter case no incentive to create new knowledge will exist. Thus a greater diffusion could have a higher economic growth impact, but at the same time the inexistence of a clear incentive could also reduce growth enhancement. This issue is only one of (the) several extant trade-off debates concerning the IPR-economic growth relationship.

The main purpose of this essay is to construct a survey of the theoretical and empirical literature on the relationship between Intellectual Property Rights (IPR), technological change and economic growth, as well as to expose some of the gaps in the current research. The relevance of this task is directly related to the ambiguous role that the literature has identified relating to the relationship between these dimensions. After systematization of the relevant theoretical literature, we focus on the empirical studies concerning the effect of IPR protection on innovation and economic growth. In presenting this overview, we intend to analyse to what extent empirical results allow (for) a consensual conclusion, faced as we are with the ambiguity of the theoretical contributions.

The present chapter is structured as follows. After a brief introduction, Section 2 presents an overview of the relationship between the economics of IPR, innovation and technological change. Section 3 focuses in detail on the relationship between endogenous economic growth and IPR from a theoretical perspective, whereas Section 4 offers an analysis of this relationship, but in empirical terms. Section 5 concludes, highlighting the main gaps that currently exist in this research agenda.

## 2. The economics of Intellectual Property Rights (IPR) and innovation: An overview

The conceptualization of IPR as a mean of protecting ideas is relatively recent. Several international agreements, such as the General Agreement on Tariffs and Trade (GATT), Trade-Related aspects of Intellectual Property Rights (TRIPS) and (the) World Intellectual Property Organization (WIPO) are examples of conventions and/or organizations connected with IPR (Senhoras, 2007).

IPR, in their various forms, play a crucial role in innovation systems.[1] Firms invest in innovation activities, find new products or new processes and increase their profits. To prevent the imitation of their innovations, firms can benefit from IPR protection. In this sense, IPR serve as an incentive for innovation, since knowledge has the characteristics of a public good (non-excludable and non-rival), and hence is easily appropriable. So in the case of IPR, the good is non-rival but becomes excludable. The significance of spillovers

---

[1] Arrow (1962) was a pioneer in addressing the economics of IPR. However, early authors such as Adam Smith, John Stuart Mill and J. W. Goethe had already conceived the patent as a price society must pay for discovery, which was fundamental for the unfettered diffusion of useful knowledge. Furthermore, it had already been recognized that the complete specification of the patent made the technological details more accessible to others (Mokry, 2009).

associated with technological knowledge being widely recognized, the related literature clearly stresses the importance of property rights, patents and other policies designed to protect innovative firms from spillovers. Nevertheless, spillovers are crucial for technology transfer and development (*e.g.*, Hall and Rosenberg, 2010). Hence, within this framework, a topic that is frequently discussed concerns the optimal patent length and the consequent trade-off between dynamic efficiency and static efficiency.

IPR play an important role not only in the innovation system but also on structural dynamics across sectors and countries, and over time. Authors such as Langford (1997) conclude that, despite there being some disadvantages, one of the most important economic effects of IPR is that they induce innovation, increasing the possibilities of technology transfer.

Resources, competences and dynamic capabilities are addressed within this wider broader discussion concerning appropriability. According to Hall and Rosenberg (2010: 689), "resources are firm-specific assets that are difficult, or impossible, to imitate. They are stocks, not flows." Resources are most likely to be intangibles; they are not easily transferred, some examples being intellectual property rights and know how processes,. As regards competences, they "are a particular kind of organizational resource", since "[t]hey result from activities that are performed repetitively or quasi-repetitively" (Hall and Rosenberg, 2010: 690). So routines are closely linked to competences. The firm's resources are considered sources of advantage, and in this context IPR correspond to firm-specific, intangible resources, which are not easy to transfer to other firms because it is difficult or even impossible to imitate them.

According to Mokyr (2009) and others, it is important to pay attention to the difference between institutions which stimulate technological progress and institutions that support the growth of markets by protecting property rights. In a completely unlegislated society, technological progress is less likely. Yet in order for rapid technological change to occur, it is necessary to eliminate some property rights. So this author is forced to ask "What kind of institutions encouraged technological progress?" (Mokyr, 2009: 349). He starts by emphasising the idea that incentives are a requirement for inventions and IPR offer incentives for successful inventors. Using the historical fact that the number of patents was stagnant until the mid- 18th century, and suddenly started growing in the 1750s, the author concludes that IPR show how institutions contributed to the origins of the Industrial Revolution. However, Mokyr also states that the main difficulty lies not in whether the patent system has a positive effect on technological progress in equilibrium but whether the effect could be sufficiently large to explain a considerable share of the acceleration in technological progress that it is intended to explain. Furthermore, it is interesting to know whether other institutions could have been similar to or even more important than the patent system. Mokyr (2009) concludes that, even as far as the historical importance of patents on the Industrial Revolution is concerned, the impact is not clear.

Cozzi (2009) discusses the possibility of innovation and growth without IPR. His main [line of] reasoning is that, since the main engine of economic growth is innovation, IPR may not necessarily be crucial for innovation and growth. In other words, although the IPR regime allows innovators to be rewarded for their innovations, constituting a mechanism whereby they are stimulated to innovate , innovation is still possible in the absence of IPR through other means, such as education (see also Greennhalgh and Rogers, 2007). Furukawa (2007) and Horii and Iwaisako (2007) also show that increasing patent protection against imitation has ambiguous effects on R&D and growth.

Hence, we may question whether the rise in profits associated with a patent increases the incentive to innovate. Initially the answer would be that two incentives are better than just one (Cozzi, 2009). However, as Cozzi (2009) also mentions, Haruyama (2009) proves that this is not always the case, because in a very populated world, the introduction of IPR could have adverse effects on the skill premium, which could consequently lead to a reduction in the tacit knowledge incentive and intensify the expected capital loss resulting from obsolescence.

The issue of appropriability is of course related to "profiting from the innovation framework" (Hall and Rosenberg, 2010: 698) and to Schumpeter's concept of creative destruction. To guarantee profits from innovation efforts and to protect inventors/innovators from imitators, two possibilities are presented: strong natural protection and strong intellectual property protection, both of which are related to appropriability regimes (Hall and Rosenberg, 2010). Patents can also be a means of protecting inventors/innovators from their rivals and ensuring the generation of profits. However, the use of patents is considered imperfect because "they are especially ineffective at protecting process innovation". This ineffectiveness is associated with e.g. the existence of considerable legal and financial requirements to prove they have been violated or with the presence of weak law enforcement relating to intellectual property (Hall and Rosenberg, 2010: 700). Thus patents act as an incentive to innovate while at the same time possibly discouraging some innovators and therefore reducing knowledge spillovers (Panagopoulos, 2009). Therefore, a concave relationship between patent protection and innovation may emerge, differing from the relationship advocated by Arrow (1962), which argues that stronger patent protection brings about leads to more innovations.

In brief, some authors criticize the argument that strong patent protection offers greater incentives to innovators and therefore increases economic performance. Cohen et al. (2000) maintain that the increasing number of patents is not necessarily a sign of their greater effectiveness. Both empirical contributions such as those of Hall and Ziedonis (2001) and theoretical approaches such as those of O'Donaghue et al. (1988) lend support to this latter perspective. Moreover, as also stressed by Panagopoulos (2009), Horii and Iwaisako (2005) maintain that stronger intellectual property protection reduces the number of competitive sectors. Since it is easier to innovate in these sectors than in monopolistic sectors, this study advocates that the innovator tends to be concentrated in a smaller number of competitive sectors.

Chu (2009b) studies the effects of IPR on the specific framework of macroeconomics. He stresses that since it is not possible to meet or recreate ideal situations in the real world, market failures can engender the overprovision or underprovision of certain resources. In fact, whereas the competitive market or Walrasian equilibriums are efficient, leading to the Pareto efficient allocation of resources, competitive conditions are difficult to come by in real economies. For example, investment in R&D activity has two implications in terms of returns: the social return and the private return. Empirical studies in this area (*e.g.*, Jones and Williams, 1988, 2000) show that the social return to R&D is much higher than the private return. This being the case, R&D, innovation, economic growth and social welfare would increase towards the socially optimal level were market failure to be overcome.

Within this context, Chu (2009b) stresses the relevance of quantitative dynamic general-equilibrium (DGE) analyses for studying the macroeconomic repercussions of rising IPR protection. He further emphasises that, although some empirical evidence points to a positive relationship between IPR protection and innovation, this evidence appears to be stronger in the case of developed rather than developing countries. Hence, this author maintains that the optimal level of patent protection[2] leads to a trade-off between the social benefits of improved innovation and the social costs of multiple distortions and income inequality. In an open economy, achieving the globally optimal level of protection demands international coordination rather than the harmonization of IPR protection.

Another interesting question in terms of policy implications is the magnitude of welfare gains from changing the patent length towards its socially optimal level. Kwan and Lai (2003) found that the extension of a patent's effective lifetime would lead to a significant increase in R&D and welfare. But Chu (2009a) maintains that while the extension of patent length beyond 20 years leads to a negligible increase in R&D and consumption, the limitation of the patent length leads to their significant reduction. So it seems that patent length is not an effective instrument for increasing R&D in most industries. In line with this argument, patent reform in the USA implemented in the 1980s focused on other aspects of patent rights such as patentability requirements (the invention would have to be new and non-obvious). Nevertheless, O'Donoghue and Zweimuller (2004) also show that if the patentability requirements are lowered, there will be contrasting effects on R&D and innovation. On the one hand, it becomes easier for an inventor to obtain a patent, which increases the R&D incentives. On the other hand, the amount of profits generated by an invention would decrease due to its smaller quality improvement, so the possibility that the next invention is patentable takes away market share from the current invention, decreasing R&D incentives. The policy implication mentioned by Chu (2009b) is the ambiguous effect of lowering the patentability requirement on R&D and growth.

Another instrument also discussed in Chu (2009b) is the patent breadth (the broadness or the scope of a patent) that determines the level of patent protection for an invention against imitation and subsequent innovations. There are two types of patent protection: the lagging breadth and the leading breadth. In relation to the former, Li (2001), using the Grossman and Helpman (1991) model, found a positive effect of the lagging breadth on R&D and growth; i.e. the increase in protection against imitation improves the incentives for R&D. This unambiguous positive effect emerges because larger lagging breadth allows (the) monopolists to charge a higher markup (Li, 2001).

Chin (2007), Furukawa (2007) and Horii and Iwaisako (2007) also show that the increase in patent protection against imitation exerts ambiguous effects on R&D and growth. Chu (2009b), basing his hypothesis on these three works, concludes that if IPR protection has asymmetric effects on different generations of households, it can also have a negative effect on innovation. Leading breadth is also discussed, underlining the point that increasing leading has opposite effects on the incentives for R&D. Once more, Chu (2009b) reports on

---

[2] Relative to patent policy, there are some instruments that can be used to influence the incentives to R&D and innovation, to the extent that they will affect economic growth. One example of these instruments is the patent length that establishes the statutory term of patent. Judd (1985) cited in Chu (2009b) argues that the optimal patent length is infinite, whereas Futagami and Iwaisako (2003, 2007) maintain, in a version of the Romer model, that the optimal patent length is finite.

O'Donoghue and Zweimuller (2004) and their analysis of Grossman and Helpman's (1991) model, to show the following: while the profits generated by an invention increase due to the consolidation of market power through generations of inventors, leading to a positive effect on R&D, the delayed rewards from profit sharing occasion a lower present value of profits received by an inventor, thus bringing about a negative effect (the profit growth rate is lower than the interest rate). This negative effect is also known as blocking patents (Chu, 2009b).

To sum up, we can conclude from the analysis of the different studies discussed above that the relationship between IPR and economic performance is ambiguous.

Although the codification of patents and copyright laws, as well as the regulation of privileges, emerged in the late 15th century, the concern with the relationship between IPR and economic growth only began in the 20th century, gathering pace as time went on. The first really relevant studies regarding IPR and growth emerged around the 1980s or even 1990s,[3] which corresponds with the emergence of the New Economic Growth Theory, also known as [the] Endogenous Economic Growth Theory, in the 1980s (Romer, 1994). The next section discusses and compares these two issues.

### 3. The bridge between IPR and Endogenous Economic Growth: Main theoretical contributions[4]

Innovation has assumed increasing importance in economic growth theory. In this context, it is consensually recognized as a crucial engine of growth (for example, Romer, 1990; Aghion and Howitt, 1992), and many studies have discussed the role of knowledge and technology in growth and development (*e.g.*, Hall and Rosenberg, 2010).

In particular, some authors focus their attention on the relation between IPR and growth. For instance, Dinopoulos and Segerstrom (2010) develop a model of North-South trade with multinational firms and economic growth in order to formally evaluate the effects of stronger IPR protection in developing countries. These effects have been the subject of intense debate, with one side advocating stronger IPR protection reform[5]and the other opposed to this (Taylor, 1994).

The former view argues that the reform would promote innovation and benefit developing countries because it would contribute to more rapid economic growth and would accelerate the transfer of technology from developed to developing countries. The latter argues that stronger IPR protection would neither accelerate economic growth nor transfer international

---

[3] Towse and Holzhauer (2002) have compiled a selection of the most important articles relating to the economics of intellectual property, and show that, in essence, they are of 20th century provenance, belonging in particular to the 1980s and 1990s. Additionally, on 29th September 2011, in a piece of internet research conducted in "SCOPUS", using "Intellectual Property Rights" and "endogenous growth model" as search words (in all text) and collecting only journal articles (including reviews), we obtained 56 records, the first dating from 1991.

[4] The selection of these studies was based on Towse and Holzhauer (2002), Pejovich (2001), Cantwell (2006) and on a thorough search of related literature on several international bibliographic databases, including Econlit and Scopus.

[5] This reform emerged from the Uruguay Round in 1994, more specifically from the TRIPS agreement, whose aim was to establish minimum standards of IPR protection by all WTO members up to 2006.

technology more quickly, since it only "results in the transfer of rents to multinational corporate patent holders headquartered in the world's most advanced countries especially in the US" (Dinopoulos and Segerstrom, 2010: 13).

Dinopoulos and Segerstrom (2010) also offer an overview of several contributions focusing on multinationals and relating to this issue. Glass and Saggi (2002), Sener (2006) and Glass and Wu (2007) show an unambiguous relation(ship) between strong IPR protection in the South and a lower rate of technology transfer, while Helpman (1993), Lai (1998), Branstetter et al. (2006) and Branstetter et al. (2007) reach the opposite conclusion. However, it is worth mentioning that, in all those previous models, the absence of R&D spending by affiliates, which is not empirically sustained (Dinopoulos and Segerstrom, 2010) is assumed. Dinopoulos and Segerstrom (2010: 14), in an effort to be coherent in considering this empirical evidence, consider that "R&D conducted by the affiliates in developing countries is focused on the absorption of patent-firm technology and on its modification for local markets." This study finds a positive relationship between stronger IPR in the South and a permanent increase in the rate of technology transfer from the North to the South. Additionally, this strong protection in the South results in a temporary increase in the Northern innovation rate and in a permanent decrease in the North-South wage gap. Hence, Dinopoulos and Segerstrom (2010) conclude that, under these conditions, Southern strong IPR protection promotes innovation in the global economy and this explains the faster growth of several developing countries compared with the growth performance of typical developed countries.

Moreover, Dinopoulos and Segerstrom (2010) analyze the long-term welfare effects, and at this level some contradictions emerge. In some North-South trade models, such as those proposed by Lai (1998), Branstetter et al. (2007) and Glass and Wu (2007), patent reform increases the economic growth rate permanently (and therefore the consumers must be better than they would be without patent reform). In other models, such as in Glass and Saggi (2002) and Sener (2006), patent reform permanently decreases the economic growth rate (and consequently consumers must be worse). In Dinopoulos and Segerstrom's (2010) model, growth is semi-endogenous and so the long-term welfare effects are ambiguous, because patent reform does not permanently alter the economic growth rate. Nevertheless, by combining all the effects gleaned from the related literature, the authors find optimistic long-term welfare effects in those developing countries with strong IPR protection. Moreover, as regards the two possible ways of transferring technology between two countries, FDI (Foreign Direct Investment) and imitation, Dinopoulos and Segerstrom (2010: 15) argue that "the effects of stronger IPR protection would depend on how important each mode of technology transfer is."

Regarding IPR protection in an open economy, Chu (2009b) emphasises three main results derived from Lai and Qui (2003) and Grossman and Lai (2004). The first indicates that, due to the asymmetries in terms of innovation capability, developed/northern countries would [tend to] choose a higher level of IPR protection than developing/southern countries. The second underlines the fact that if the North's level of IPR protection, such as TRIPS, were imposed on (the) southern countries, it would lead to a welfare gain (loss) in the North (South). And finally, although TRIPS require the harmonization of IPR protection, this harmonization is neither necessary nor sufficient for the maximization of global welfare.

Chu and Peng (2009), quoted by Chu (2009b), also consider the effects of IPR protection on income inequality across countries and find that stronger patent rights in one country tend to lead to an increase in economic growth and income inequality both in domestic and foreign countries. Another result of this research is that TRIPS tend to improve or reduce global welfare according to the domestic importance of foreign goods. Thus, only if these goods were sufficiently important for domestic consumption would the harmonization of IPR protection that the TRIPS require improve global welfare.

Cozzi (2009) also highlights the role of IPR in economic growth in both developed and developing countries. Typically, while the developed countries are the northern countries, which create new varieties of goods and services, the developing countries are the southern countries, which have a production cost advantage. In this sense, the source of growth is the horizontal innovation of new intermediate products. In the case of northern firms, they may export (the) intermediate goods, they may directly invest in the South (through knowledge transfer), or they may grant a licence for their product (complete transfer). These firms desire to transfer the maximum possible knowledge, but this implies the transfer of more knowledge about their patented goods. The southern firms can try to undertake costly imitation activities, so that in the South IPR protection is not complete: the more intensive is the knowledge transfer, the higher is the probability of southern firms imitating their northern counterparts. The greater is the IPR protection, the higher is the equilibrium FDI,[6] which makes it possible to improve the international division of labour. Thus, while very high IPR protection implies licensing - this method being the most efficient - very low IPR protection induces the firms of the North not to transfer at all, but to produce domestically and to export their intermediate goods to the South. The advantage of this last situation is the absence of unproductive Southern imitation costs. Cozzi (2009) also maintains that different IPR effects can exist: the combination of the general equilibrium effects of adverse incentives and wasteful imitation costs implies that the increase in international IPR protection is beneficial to the welfare of the South if the initial level of IPR is already above a certain threshold. However, in the case of weak protection of initial IPR, the increase in protection might be dangerous for (the) southern consumers.

Globalization, inequality and innovation are phenomena crucially associated with IPR. Spinesi (2009) extended Dinopoulos and Segerstrom's (1999) work on Schumpeterian economic growth, by studying the relations between all those dimensions. Among others issues, Spinesi emphasises that IPR achieve a similar result even in the presence of constant returns to scale. This result is advantageous because it would also apply in the case of firms competing *a la* Bertrand. Moreover, he finds that, while horizontal innovation has a positive level effect, it is vertical innovation that sustains the growth effect.

Panagopoulos (2009) explores the relationship between patent breadth and growth, by studying how patent breadth affects innovation and output. This study finds an inverse U relationship between patent protection and growth.

From the different studies mentioned above, we conclude that there is no consensus regarding the relationship between IPR and economic growth, including within the specific theoretical framework of endogenous growth literature. In Table 1 we offer a systematization of this theoretical literature.

---

[6] The relationship between IPR protection and FDI is also analyzed by Chu (2009b): technological transfer between northern and southern firms occurs to a significant extent *via* FDI.

| IPR conceptualization | Author (date) | Net final effect |
|---|---|---|
| Patents | Scherer (1997); Koléda (2008) | On innovation: 0 |
| | Tandon (1982); David and Olsen (1992) | On welfare: + |
| | Merges and Nelson (1994) | On technological progress: - |
| | Taylor (1994) | On economic growth: + (symmetric protection) and – (asymmetric protection). |
| | Michel and Nyssen (1998); Goh and Oliver (2002); Iwaisako and Futagami (2003) | On economic growth: + |
| | Futagami and Iwaisako (2007) | On economic growth: + (finite patent length) and – (infinite patent length); a patent strategy with a finite patent length is optimal. |
| | Naghavi (2007) | On South welfare: + (if attract foreign investment in less R&D intensive industries or if they stimulate innovation in high technology sectors). |
| | Dinopoulos and Kottaridi (2008) | On economic growth and on income distribution: + (if each country selects the level of patent enforcement optimally, with the North having an incentive to choose stronger IPR protection than the South). |
| | Eicher and Garcia-Peñalosa (2008); Chu (2009a) | On economic growth: 0 |
| | Panagopoulos (2009) | On economic growth: a concave relationship. |
| Index of Patent Rights from Park (2008) | Chu (2010) | On economic growth: +; on income inequality: 0 |
| Patent length and breadth; protection trademarks; copyrights and trade secrets; and the degree of enforcement. | Kwan and Lai (2003) | On economic growth: optimal degree of IPR protection. |
| Copyright | Novos and Waldman (1984) | On social welfare: + |

| IPR conceptualization | Author (date) | Net final effect |
|---|---|---|
| Patent length and breadth; copyright policy | Landes and Posner (1989) | On welfare associated with a given work: - |
| | Furukawa (2007) | On economic growth: - (when the impact of accumulated experience on productivity is large enough, an inverted U relationship is suggested). |
| Increase in imitation costs | Stryszowski (2006) | On economic growth in technologically lagging countries: 0 |
| | Glass and Saggi (2002); Mondal and Gupta (2008) | On innovation and on FDI: - |
| | Mondal and Gupta (2009); Connolly and Valderrama (2005) | On welfare: + (both in North and in South, although the marginal welfare gain is higher in the former than in the latter) |
| | Wu (2010) | On innovation: + |
| Tariffs; Increase in the costs of imitation | Datta and Mohtadi (2006) | On South's economic growth: tariffs (-); IPR (-) |
| Imitation intensity | Mondal and Gupta (2006); Glass and Wu (2007); Zhou (2009) | On innovation: 0 |
| | Dinopoulos and Segerstrom (2010) | On innovation: + |
| Imitation probability and the return of innovation. | Horii and Iwaisako (2007) | On economic growth: 0 |
| Royalties | Saint –Paul (2008) | On welfare: + |
| N/a | Furukawa (2010) | On innovation: inverted U |

N/a: not applicable; 0: ambiguous or inconclusive net effect; +(-): positive (negative) net effect.
*Own elaboration.*

Table 1. The impact of IPR on innovation and growth: a synthesis of the theoretical literature

As mentioned above, the relevant theoretical literature points to both positive and negative effects of patent protection on innovation (Chu, 2009b). For example, Furukawa (2007) and Eicher and Garcia-Peñalosa (2008) refute the idea that stronger IPR protection is always better. Using an endogenous growth model with costless imitation, Furukawa (2007) proves that IPR protection cannot increase economic growth, whereas Eicher and Garcia-Peñalosa (2008) support the idea that the relationship between IPR and economic growth is ambiguous. Iwaisako and Futagami (2003), Mondal and Gupta (2006) and Futagami and Iwaisako (2007) identify two opposite effects on this relationship. Wu (2010) presents inconclusive results that depend on such features as the countries' level of development or the channel of technology transfer. Scherer (1977) also maintains that patents involve an impact that depends on such factors as the market position of the innovator, the features of the technology (whether it is easy or difficult for it to be imitated), the cost, the risks and the potential payoffs from innovation. Furukawa (2010) and Panagopoulos (2009) find an inverted U relationship between IPR protection and innovation (and economic growth). Kwan and Lai (2003) and Connolly and Valderrama (2005) argue that IPR are important to R&D investment and (to) welfare.

Table 1 is also helpful in showing that different authors use different concepts of IPR. Some of them (Scherer, 1997; Tandon, 1982; David and Olsen, 1992; Merges and Nelson, 1994; Taylor, 1994; Michel and Nyssen, 1998; Goh and Oliver, 2002; Iwaisako and Futagami, 2003; Futagami and Iwaisako, 2007; Naghavi, 2007; Dinopoulos and Kottaridi, 2008; Naghavi, 2007; Dinopoulos and Kottaridi, 2008; Eicher and Garcia-Peñalosa, 2008; Koléda, 2008; Chu, 2009a; Panagopoulos, 2009) limit the definition of IPR to one of their forms – patents (considered as the most important form of IPR, as discussed above). Others use distinct definitions, e.g. Glass and Saggi (2002) and Mondal and Gupta (2008, 2009), who define IPR as the rise in the imitation cost. Connolly and Valderrama (2005) give a similar definition, assuming that imitators pay a licence fee which is similar to an increase in the fixed cost of the imitative research; Kwan and Lai (2003) consider IPR part of the imitation rate which can be influenced by some factors such as patents, trademarks, copyrights and trade secrets; Furukawa (2007) also defines IPR as a mixed measure of patent and copyright; Glass and Wu (2007) associate the measure of IPR with (the) imitation intensity, whereas Dinopoulos and Segerstrom (2010) define IPR as a reduction in the exogenous rate of imitation.

Despite these different ways of defining IPR, we do not find evidence of significant differences in terms of the results obtained. In fact, we have two studies in the table that achieve the same results using different measures of IPR: Furukawa (2007) and Panagopoulos (2009). Both suggest an inverted U relationship between IPR and economic growth, although the former defines IPR as a mix of patent and copyright measures, while the latter defines IPR only as patents. Furukawa (2010) also finds the same relationship, although he does not define IPR.

Two of the articles in Table 1, Stryszowski (2006) and Mondal and Gupta (2008), compare their assumptions and/or conclusions with other studies – some of them also analyzed in the present work. Stryszowski (2006) identifies and discusses studies which maintain that strong IPR protection is beneficial for (the) innovating economies (e.g., Connolly and Valderrama, 2005). However, this study also highlights works that have found negative effects of IPR protection on lagging economies, based on the existence of a mechanism in which strong IPR protection tends to raise consumer prices and to diminish trade benefits that could be essential for developing economies (for example, Hekpman, 1993). Mondal and Gupta (2008) discuss several studies based on their distinct assumptions concerning the innovation framework (quality ladder framework *versus* product variety framework), and the alternative ways of treating imitation and of strengthening IPR protection, etc. Following this they present the assumptions of their own model, characterized by the use of a product variety model, a North-South model with endogenous innovation, imitation and multinationalisation, where innovation activities are set as costly and there is an endogenous rate of imitation. Lai (1998) gives a close approximation to this latter one, except for two features: the endogenous imitation rate in the South, given that imitation is considered costly; and the introduction of two kinds of labour in the South - skilled and unskilled. In line with these two distinct assumptions, the results achieved are also different from Mondal and Gupta's (2008).

To sum up, we can state that the studies presented in Table 1 do not show a pattern regarding the relationship between IPR and economic growth (*via* welfare, for instance). From this table we can also state that patents are the most widely used measure of IPR in theoretical works. Hence, at this stage, we state the existence of two main gaps in this literature: the scarcity of studies, so that a potential research line would be to dig more deeply in this field; and the excessive focus on patents as an IPR measure, which neglects the potential impact of other instruments, such as copyrights, which are crucial for the development of specific ICT industries such as information technology and software.

In the next section we develop an analysis of the empirical studies concerning this same relation(ship) between IPR and endogenous economic growth.

## 4. IPR and Endogenous Economic Growth: Where do we stand? Insights from the empirical literature

After the systematization of the relevant theoretical literature in the previous section, we focus on empirical studies into the effect of IPR protection on innovation and on economic growth. In this review (*cf.* Table 2), we intend to show whether the empirical results permit us to reach a sustainable conclusion, faced as we are with the confirmed ambiguity of the theoretical contributions.

As we have seen above, according to the theoretical literature, patent protection generally has positive and negative effects on innovation (*e.g.*, Chu, 2009b). However, empirical studies usually find a positive effect, which according to Chu (2009b) is explained by the domination of the positive effects over the negative ones. As we can see in Table 2, the empirical evidence suggests a positive relationship between IPR protection and innovation, although with some 'restrictions' in the sample. For instance, the positive result is true only for developing countries (Falvey *et al.*, 2009; Chen and Puttitanun, 2005). At a first glance, we could expect the opposite result. However, while some works, for example Park (2005) and Kanwar and Evenson (2003), generally find a positive effect, Chen and Puttitanun (2005) explain that, on the one hand, lower IPR can facilitate imitation, while on the other hand, innovation in developing countries increases in proportion to greater IPR protection. Moreover, these authors state that the optimal degree of IPR protection may depend on the country's development level. Furthermore, Falvey *et al.* (2006) find evidence of a positive effect between IPR and economic growth for both low and high-income countries, but not for middle-income countries. According to the latter, the positive relation between IPR and economic growth in low-income countries cannot be explained by the potential fostering of R&D and innovation, but by the idea that stronger IPR protection promotes imports and inner FDI from high-income countries without negatively affecting the national industry based on imitation.

Hence, when the division between developed and developing countries is considered, the effects of patent rights on R&D are rendered ambiguous: for instance, according to Chen and Puttitanun (2005), in developing countries there is a positive and significant relationship between IPR protection and innovation, while according to Park (2005), there is an insignificant effect of IPR protection on R&D.

| Measure of IPR | Methodology | Sample | Author (date) | Net estimated effect |
|---|---|---|---|---|
| Park and Ginarte (1997) Index of Patent Rights[7] | Econometric analysis – Seemingly Unrelated Regressions (SUR) | Cross-section of countries for the period 1960-1990. | Park and Ginarte (1997) | + |
| | Econometric analysis – cross section | 48 countries for the period 1980 and 2000 (Sources: World Intellectual Property Organization (WIPO) and Penn World Table 6.1) | Xu and Chiang (2005) | |
| | Econometric analysis – panel data | 64 developing countries over the 1975– 2000 period (Sources: World Development Indicators and Statistical Yearbook by UNESCO (UNESCO, 1995, 1997, 2000); patent data come from the United States Patent and Trademark Office Website) | Chen and Puttitanun (2005) | |
| | | 79 countries and four sub-periods: 1975-79, 1980-84, 1985-89 and 1990-94. | Falvey et al. (2006) | + (for low-income and high-income countries) |
| | | 80 countries for the period 1970–1995. (Sources: PennWorld TableMark 6.1, updated version of Summers and Heston, 1991; UNCTAD, 2005; World Bank, 2005; Ginarte and Park, 1995; Easterly and Sewadeh, 2005; Hall and Jones, 1999; and Barro and Lee, 2000). | Groizard (2009) | **Ambiguous:** + (FDI is higher for countries with stronger intellectual property protection). - (Negative relationship between IPR and human capital indicators). |

[7] This index is a simple sum of the scores attributed to each of the five categories of patent rights (score from 0 to 1) on a scale of 0 to 5, with a larger number indicating stronger patent rights. The five IPR categories are the patent duration, the coverage, the enforcement mechanisms, the restrictions on patent scope and the membership in international treaties.

| Measure of IPR | Methodology | Sample | Author (date) | Net estimated effect |
|---|---|---|---|---|
| | | 69 developed and developing countries over the period 1970–1999 (Sources: World Bank's World Development Indicators, 2001, Jon Haveman website, OECD's International Trade by Commodity Statistic (Historical Series, 1961–1990), International Trade by Commodity Statistic (1990–1999), and Barro and Lee (2001) database. | Falvey *et al.* (2009) | **Non-linear:** (Depends on level of development, the imitative ability and the market size of the importing country). |
| Park and Ginarte (1997) Index extended by Park (2008a) | | 50 countries (Sources: Ginarte and Park, 1997; Sachs and Warner, 1995; Hofstede, 1984 and UNESCO, 1998). | Varsakelis (2001) | |
| | | 32 countries for the period between 1981 and 1990 (Sources: Ginart and Park, 1997; Esty *et al.*, 1998; United Nations, 1999; Word Bank, 2000; Barro and Lee, 2000; Heston *et al.*, 2001 and Pick's Currency Yearbook and World Currency Yearbook, several years). | Kanwar and Evenson (2003) | |
| Eight indexes:[8] index of patent rights constructed from Ginart and Park (1997) and Park and Wagh (2002); index of copyrights; index of trade-marks; index of parallel import protection; index of software rights; index of piracy rates; index of enforcement provisions and index of enforcement in practice. | Econometric analysis | 41 countries (Sources: Penn World Tables (Version 5.6a), World Bank Development Indicators and UNESCO's Statistical Yearbook). | Park (2005) | + |

[8] For the first three indexes (relative to patents, copyrights and trade-marks) the index consists of four sub-categories: coverage, duration, restrictions and membership in international treaties. Enforcement can also be included as a sub-category (such as in Ginarte and Park, 1997) but it was considered useful to separate this sub-category and treat it as another index.

| Measure of IPR | Methodology | Sample | Author (date) | Net estimated effect |
|---|---|---|---|---|
| Patent rights index data (Park, 2001) | Semiparametric model | 21 countries for the period 1981 and 1997 (Sources: World Bank World Development Indicators, 1999; and UNESCO). | Alvi *et al.* (2007) | |
| Patents | Econometric analysis | Firms in the chemical, drug, electronics and machinery industries | Mansfield *et al* (1981) | |
| | | Japanese and U. S. patent data on 307 Japanese firms (Sources: Japan Development Bank Corporate Finance Database, Kaisha Shiki Ho R&D, JAPIO, CASSIS CD-ROM, RAI patent database and Hoshi and Kashyap, 1990). | Sakakibara and Branstetter (2001) | 0 |
| | | 4 countries (manufacturing sector divided into 12 subgroups) between 1990 and 2001 (Sources: OECD STAN, EPO and PERINORM). | Blind and Jungmittag (2008) | |
| Impact of patent reform[9] | | 16 countries over the 1982-1999 period (Sources: U.S. Bureau of Economic Analysis (BEA) Survey; World Intellectual Property Rights Organization (WIPO)). | Branstetter *et al.* (2005) | + |
| N/a (property rights) | | 68 developed and developing countries between 1976 and 1985 (Sources: World Development Report 1988, Summers and Heston, 1988; World Bank, 1990; and Scully and Slottje, 1991). | Torstensson (1994) | |

N/a: not applicable; 0: ambiguous or inconclusive net effect; +(-): positive (negative) net effect.
*Own elaboration.*

Table 2. The impact of IPR on innovation and growth: a synthesis of the empirical literature

Chu (2009b), in giving a plausible explanation for this contrast emanating from empirical analyses, points to the fact that developed countries are typically close to the technology frontier, and that consequently economic growth in these countries requires original innovations, while developing countries are normally further away from the technology frontier, thus enabling economic growth to be driven by the reverse engineering of foreign technologies. Therefore, stronger patent rights, which discourage the reverse engineering of foreign technologies, can asphyxiate the innovation process in developing countries. Chu

---

[9] "Each reform can be classified according to whether or not it expanded or strengthened patent rights along five dimensions: 1) an expansion in the range of goods eligible for patent protection, 2) an expansion in the effective scope of patent protection, 3) an increase in the length of patent protection, 4) an improvement in the enforcement of patent rights, and 5) an improvement in the administration of the patent system." (Branstetter *et al.*, 2006: 14).

(2009b) emphasises that the increase in the level of patent protection by policymakers is similar to giving more market power to monopolists, which intensifies the deadweight loss. He recalls Nordhaus' (1969) contribution in stating that the optimal level of patent protection should trade-off the harmful effects of IPR protection on society, even when stronger patent rights are growth-enhancing, against the welfare gain from innovation. Hence, distortionary effects of IPR protection could emerge. The latter author also emphasizes that, when skilled and unskilled workers are assumed, (the) strong patent protection increases the return to R&D and the wage of R&D workers.

Through analysing the net effect of the IPR on economic growth we can state that it is not easy unequivocally to draw conclusions regarding the sign of that effect, despite the prevalence of the positive sign (cf. Table 2). We find evidence of both a positive sign and a negative sign. Possible explanations, beyond the focus on a patent index for measuring IPR (as also highlighted by Chu, 2009b, which mentions that it is not clear how each type of patent rights influences innovation on empirical grounds), are: the fact that some studies do not analyze the direct effect between IPR and economic growth; the adoption of different methodologies and of distinct samples. Hence, the gaps already mentioned when discussing the theoretical contributions clearly emerge here in association with the empirical studies. Once again, insufficient analysis, even more striking at the empirical level, and the excessive focus on patents as means of IPR measurement are evident.

## 5. Conclusion

This study supports the conclusion that there is no clear relationship between IPR and economic growth. Theoretical literature indicates that IPR protection has positive, negative or even ambiguous (or inconclusive) effects on innovation.

After a thorough review of this theoretical literature it has been possible to identify some gaps in the research agenda. Firstly, in general, this research does not study the direct and net effect of IPR on economic growth. In fact it only analyzes the relationship between IPR-induced factors and economic growth, or the impact of IPR on other economic indicators such as welfare, technological change, FDI, R&D, innovation, etc.. This happens because a standard argumentation is adopted, maintaining a strict relation between these elements and economic growth. For instance, Mondal and Gupta (2006: 27) point out that "[t]echnological change plays the most important role in determining a country's rate of economic growth. Strengthening the Intellectual Property Rights (IPR) is an important factor that motivates technological change". Furthermore, Koléda (2002: 1) argues that "[i]nnovation is an important source of economic growth". Mansfield (1986: 173) holds that "[t]he patent system is at the heart of our nation's policies toward technological innovation." Secondly, there is a disproportionate focus on patent measurement as a proxy for IPR, and thirdly, it is clear that there is a scarcity of studies in this field , particularly in empirical terms.

Despite the divergence of results regarding theoretical studies, most empirical studies find a net positive effect, which means that positive effects of IPR protection outweigh the negative effects. A possible explanation for this is that the empirical measure of patent protection, which is typically used, is just a summary of the statistics relating to the different categories of patent rights and so it is not clear how each type of patent rights influences innovation on empirical grounds (Chu, 2009b).

From the above, we consider that more research on this specific topic is crucial in order to further advance our understanding of the relation(ship) between IPR and economic growth on a worldwide scale, and to be able clearly to go beyond the strict modelling frame.

## 6. References

Arrow, K. (1962), "Economic welfare and the allocation of resources for invention", in *National Bureau of Economic Research Volume, The Rate and Direction of Inventive Activity*, Princeton: Princeton University Press, 609-625

Aghion, P. and Howitt, P. (1992), 'A model of growth through creative destruction', *Econometrica*, 60, 323-351.

Alvi, E., Mukherjee, D. and Eid, A. (2007), 'Do patent protection and technology transfer facilitate R&D in developed and emerging countries? A semiparametric study', *Atlantic Economic Journal*, 35(2), 217-231.

Besen, S. M. and Raskinind, L. J. (1991), 'An introduction to the law and economics of intellectual property', *Journal of Economic Perspectives*, 5 (1), 3-27.

Branstetter, L., Fisman, R. and Foley, C. F. (2005), 'Do stronger intellectual property rights increase international technology transfer? Empirical evidence from U.S. firm-level data', *National Bureau of Economic Research*, Working Paper No. 11516.

Cantwell, J. (2006), '*The Economics of Patents*', The International Library of Critical Writings in Economics.

Chen, Y. and Puttitanun, T. (2005), 'Intellectual property rights and innovation in developing countries', *Journal of Development Economics*, 78, 474-493.

Chu, A. C. (2009a), 'Effects of patent length on R&D: a quantitative DGE analysis', *Journal of Economics*, 14, 55–78.

Chu, A. C. (2009b), 'Macroeconomic effects of intellectual property rights: a survey', *Academia Economic Papers*, 37(3), 283-303.

Chu, A. C. (2010a), 'Effects of patent length on R&D: a quantitative DGE analysis', *Journal of Economics*, 99, 117-140.

Chu, A. C. (2010b), 'Effects of patent policy on income and consumption inequality in a R&D growth model', *Southern Economic Journal*, 77(2), 336-350.

Connolly, M. and Valderrama, D. (2005), 'North-South technological diffusion: a new case for dynamic gains from trade', *American Economic Review*, 95 (2), 318-322.

Cozzi, G. (2009), 'Intellectual Property, innovation, and growth: introduction to the special issue', *Scottish Journal of Political Economy*, 56(4), 383-389.

Datta, A. and Mohtadi, H. (2006), 'Endogenous imitation and technology absorption in a model of North-South trade', *International Economic Journal*, 20(4), 431-459.

Dinopoulos, E. and Kottaridi, C.(2008), 'The growth effects of national patent policies', *Review of International Economics*, 16(3), 499-515.

Dinopoulos, E. and Segerstrom, P. (1999), 'A Schumpeterian model of protection and relative wages', *American Economic Review*, 89 (3), 450-472.

Dinopoulos, E. and Segerstrom, P. (2010), 'Intellectual property rights, multinational firms and economic growth', *Journal of Development Economics*, 92, 13-27.

Eicher, T. and García-Peñalosa, C. (2008), 'Endogenous strength of intellectual property rights: implications for economic development and growth', *European Economic Review*, 52, 237-258.

Falvey, R., Foster, N. and Greenaway, D. (2006), 'Intellectual property rights and economic growth', *Review of Development Economics*, 10(4), 700-719.

Falvey, R., Foster, N. and Greenaway, D. (2009), 'Trade, imitative ability and intellectual property rights', *Review World Economics*, 145(3), 373-404.

Furukawa, Y. (2007), 'The protection of intellectual property rights and endogenous growth: Is stronger always better?', *Journal of Economic Dynamic & Control*, 31, 3644-3670.

Furukawa, Y. (2010), 'Intellectual property protection and innovation: an inverted-U relationship', *Economics Letters*, 109, 99-101.

Futagami, K. and Iwaisako, T. (2007), 'Dynamic analysis of patent policy in an endogenous growth model', *Journal of Economic Theory*, 132, 306-334.

Glass, A. J. and Saggi, K. (2002), 'Intellectual property rights and foreign direct investment', *Journal of International Economics*, 56, 387-410.

Glass, A. J. and Wu, X. (2007), 'Intellectual property rights and quality improvement', *Journal of Development Economics*, 82, 393-415.

Goh, Ai- Ting and Oliver, J. (2002), 'Optimal patent protection in a two-sector economy', *International Economic Review*, 43(4), 1191-1214.

Granstrand (2005), 'Innovation and intellectual property rights', in *The Oxford Handbook of Innovation*, Oxford University Press, 266-290.

Greenhalgh, C. and Rogers, M. (2007), 'The value of intellectual property rights to firms and society', *Oxford Review of Economic Policy*, 23(4), 541-567.

Groizard, J. L. (2009), 'Technology trade', *Journal of Development Studies*, 45(9), 1526-1544.

Grossman, G. M. and Helpman, E. (1991), 'Quality ladders in the theory of growth', *Review of Economic Studies*, 58, 43-61.

Hall, B. H. and Rosenberg, N. (2010), *Handbook of the Economics of Innovation*, Elsevier: Amsterdam.

Haruyama, T. (2009), 'Competitive innovation with codified and tacit knowledge', *Scottish Journal of Political Economy*, 56(4), 390-414.

Horii, R. and Iwaisako, T. (2007), 'Economic growth with imperfect protection of intellectual property rights', *Journal of Economics*, 90 (1), 45-85.

Iwaisako, T. and Futagami, K. (2003), 'Patent policy in an endogenous growth model', *Journal of Economics*, 78(3), 239-258.

Jones, C. (1995), 'Time series tests of endogenous growth models', *The Quarterly Journal of Economics*, 110 (2), 495-525.

Jones, C. and Williams, J. (1998), 'Measuring the social return to R&D', *Quarterly Journal of Economics*, 113, 1119-1135.

Jones, C. and Williams, J. (2000), 'Too much of a good thing? The economics of investment in R&D', *Journal of Economic Growth*, 5, 65-85.

Kanwar, S. and Evenson, R. (2003), 'Does intellectual property protection spur technological change?', *Oxford Economic Papers*, 55, 235-264.

Koléda, G. (2008), 'Promoting innovation and competition with patent policy', *Journal of Evolutionary Economics*, 18, 433-453.

Kwan, Y. K. and Lai, E. L.-C (2003), 'Intellectual property rights protection and endogenous economic growth', *Journal of Economic Dynamics & Control*, 27, 853-873.

Menell, P. S. (1999), 'Intellectual Property: General Theories', entry 1600 in Bouckaert, B. and De Geest, G. (eds.), *Encyclopaedia of Law and Economics*, Cheltenham, UK and Northampton, US: Edward Elgar, 129-164.

Michel, P. and Nyssen, J. (1998), 'On knowledge diffusion, patents lifetime and innovation based endogenous growth', *Annales d'économie et statistique*, 49/50, 77-103.

Mokyr, J. (2009), 'Intellectual property rights, the individual revolution, and the beginnings of modern economic growth', *American Economic Review*, 99(2), 349-355.

Mondal, D. and Gupta, M. R. (2006), 'Product development, imitation and economic growth: a note', *Journal of International Trade & Economic Development*, 15(1), 27-48.

Mondal, D. and Gupta, M. R. (2008), 'Innovation, imitation and multinationalisation in a North-South model: a theoretical note', *Journal of Economics*, 94(1), 31-62.

Mondal, D. and Gupta, M. R. (2009), 'Endogenous imitation and endogenous growth in a North-South model: a theoretical analysis', *Journal of Macroeconomics*, 31(4), 668-684.

Naghavi, A. (2007), 'Strategic intellectual property rights policy and North-South technology transfer', *Review of World Economics*, 143(1), 55-78.

Panagopoulos, A. (2009), 'Revisiting the link between knowledge spillovers and growth: an intellectual property perspective', *Economics of Innovation and New Technology*, 18(6), 533-546.

Park, W. (2005), 'Do intellectual property rights stimulate R&D and productivity growth? Evidence from cross-national and manufacturing industries data', in J. Putnam (ed.), *Intellectual Property Rights and Innovation in the Knowledge-Based Economy*, Calgary: University of Calgary Press, 9-1 – 9-51.

Park, W. G. and Ginarte, J. C. (1997), 'Intellectual property rights and economic growth', *Contemporary Economic Policy*, Vol. XV, 51-61.

Pejovich, S. (2001), '*The Economics of Property Rights*', The International Library of Critical Writings in Economics.

Romer, P. M. (1990), 'Endogenous technological change', *Journal of Political Economy*, 98, 71-102.

Romer, P. M. (1994), 'The origins of endogenous growth', *Journal of Economic Perspectives*, 8(1), 3-22.

Saint-Paul, G. (2008), 'Welfare effects of intellectual property in a North-South model of endogenous growth with comparative advantage', *Economics (The Open-Acess, Open-Assessment E-Journal)*, 2, 1-24.

Senhoras, E. M. (2007), 'Introduction to Intellectual Property Rights in the International Relations', *Intellector Magazine*, year, III, Vol. IV (7), 1-26.

Spinesi, L. (2009), 'Intellectual property meets economic geography: globalization, inequality, and innovation strategies', *Scottish Journal of Political Economy*, 56(4), 508-542.

Stryszowski, P. K. (2006), 'Intellectual property rights, globalization and growth', *Global Economy Journal*, 6(4), article 4, 1-31.

Taylor, M. S. (1994), 'Trips, trade, and growth', *International Economic Review*, 35(2), 361-381.

Varsakelis, N. C. (2001), 'The impact of patent protection, economy openness and national culture on R&D investment: a cross-country empirical investigation', *Research Policy*, 30, 1059-1068.

Wu, H. (2010), 'Distance to frontier, intellectual rights and economic growth', *Economics of Innovation and New Technology*, 19(2), 165-183.

Xu, B. and Chiang, E. P. (2005), 'Trade, patents and international technology diffusion', *Journal of International Trade & Economic Development*, 14(1), 115-135.

Zhou, W. (2009), 'Innovation, imitation and competition', *The B. E. Journal of Economic Analysis &Policy*, 9(1), article 27, 1-14.

# 3

# The Social Consequences of Technological Change in Capitalist Societies

Tony Smith
*Iowa State University*
*USA*

## 1. Introduction

No one would dispute that modern capitalist societies have exhibited a historically unprecedented level of technological dynamism. In contrast, assessments of the social consequences of this dynamism diverge widely. It is not possible here to provide a comprehensive overview of competing assessments. I shall first present three important theoretical perspectives on this issue in normative social theory: classical liberalism, liberal egalitarianism, and Marxism. I shall then develop a historical narrative of some of the most significant social consequences of technological change in recent decades. I conclude that a Marxian framework illuminates the role of technical change in shaping the present moment of social history better than the competing frameworks considered in this paper.

## 2. Classical liberalism and the social consequences of technology

"Liberalism" is one of the most ambiguous words in our language. It has become one of the strongest terms of opprobrium used by "conservative" advocates of free markets in the United States today against their enemies, despite the fact that they are themselves part of the tradition of classical liberalism beginning with John Locke and Adam Smith.[1] For adherents of this tradition individuals are the basic unit of moral concern, and all individuals are asserted to be equally worthy of moral respect. From this *moral equality principle* it follows as a corollary that all individuals should be free to decide for themselves both their life plans and how best to carry them out. Private property rights are an essential component of a social order based on these ideas. Individuals are asserted to have a right to privately appropriate previously unowned things as well as the things they themselves make (either alone or with the aid of others who freely chose to cooperate with them). They also have the fundamental right to freely decide how to use their acquired property, including a right to undertake exchanges should they wish to do so. The exercise of these rights is, of course, subject to the constraint that the equal rights of others to do the same are acknowledged and respected.

Classical liberal theorists affirm free markets on the grounds that they institutionalize rights to liberty and property better than any feasible alternative. This rights-based argument is

---

[1] "Classical liberalism," "egalitarian liberalism," and "Marxism" are ideal types. Particular theorists grouped under a particular heading can approach the ideal type in question more or less closely. A fairly pure form of the perspective on the social consequences of technological change in capitalism developed in the following paragraphs is found in Hayek, 1976.

usually conjoined with claims about the positive consequences of free market economies, with technological dynamism at the top of the list. In these societies producers have a strong incentive to introduce process innovations enabling goods and services to be produced more efficiently, since if they lower unit costs they can lower prices and win market share. There is an equally strong incentive to introduce product innovations that allow purchasers to fulfill existing wants and needs in more satisfactory ways, or to develop new wants and needs in ways they freely choose.

The social consequences of this technological dynamism have profound normative significance within the classical liberal framework. Life plans require goods and services to be carried out. Free markets allow individuals themselves to choose the goods and services that they regard as most important to the fulfillment of their life plans. The drive to product innovations ensures that over time the goods and services that are produced necessarily tend to be those that contribute to freely chosen life plans to the greatest feasible extent. Further, the drive to process innovations ensures that over time that greatest feasible amount of those goods and services will be provided. If human flourishing can be defined as carrying out the lives we have chosen to live to the greatest feasible extent, then technological change in capitalist market societies provides the material preconditions for human flourishing.

Technological change will only have these beneficial social consequences if market competition operates effectively. This in turn requires that economic agents are confident that their property rights will be respected, that they will be able to enter into voluntary contracts providing mutual benefits, and that the terms of these contracts will be adhered to. The main role of the state in the classical liberal framework is to provide a coercive apparatus ensuring that liberty and property rights can be enjoyed in security under "the rule of law."

For libertarians the legitimate functions of the state do not go beyond measures to protect individual citizens against force and fraud. Libertarianism, however, is only one species of classical liberalism. Relatively few advocates of classical liberalism argue against government funding of basic research or basic infrastructure. Many also hold that public resources should be mobilized for training programs to ensure that the workforce possesses needed technical skills. The belief is also widely shared that environmental problems can sometimes be serious enough to warrant regulation of polluting technologies. These sorts of state policies are said to complement free markets, enhancing their technological dynamism beyond what it would otherwise be and ensuring that this dynamism contributes to human flourishing to the greatest feasible extent.

Despite support for such policies, however, there is still an important sense in which even non-libertarian classical liberals defend a "minimal state." This can be seen by considering the technology policies that classical liberalism either excludes completely or accepts to a very limited degree.

- Capital markets must be able to shift investment funds smoothly and rapidly to new innovative sectors, or to established sectors undertaking significant innovations. Burdensome state regulations of capital markets must be avoided if the positive social consequences of technological change are to be enjoyed to the fullest extent.
- Labor markets must be flexible so that labor can shift from less innovative sectors and regions to more innovative sectors and regions as smoothly and rapidly as possible. Here too state regulations impeding flexibility must be avoided.

- Restrictions on trade should also be limited in order to enable consumers to enjoy the benefits of innovative imports, and to spur domestic firms to innovate in order to compete successfully in the world market.
- Environmental regulations must not be considered prior to a complete and scientifically sound assessment of their costs and benefits. The former must not be understated, or the latter given undue weight, out of nostalgia for a romanticized and unrecoverable past. Further, the extent to which innovations can mitigate problems should not be underestimated; markets, for example, provide powerful incentives both to use costly natural resources more efficiently and to search for technological substitutes. In general, the best way to confront environmental and other difficulties is through market-driven innovation.

These arguments can easily be generalized to justify free trade and free capital flows across borders. A regime of global governance eliminating restrictions on foreign trade and investment, encouraging flexible labor markets, and avoiding unsound environmental restrictions, follows as well. If anything, these conclusions have even greater normative force on the global level. Global justice demands that individuals throughout the world have access to the material preconditions for human flourishing. Technological dynamism has proven to be the single most effective means of generating these preconditions. Global justice, classical liberals conclude, therefore demands the adaptation of measures spurring technological dynamism, and the avoidance of public policies hampering it.

## 3. Liberal egalitarianism and the social consequences of technology

Like classical liberals, liberal egalitarians take individuals as the basic unit of moral concern and affirm that all individuals are equally worthy of moral respect. They question, however, whether free markets, minimal states, and a global regime based on free trade and free flows of capital adequately institutionalize the moral equality principle. John Rawls, perhaps the most influential political philosopher of the twentieth century, insisted that this principle demands a substantial (and not merely formal) equality of basic liberty rights, fair equality of opportunity, and a distribution of income and wealth in which all citizens benefit from economic growth (Rawls, 2001). Rawls and other liberal egalitarians hold that the institutional order defended by classical liberals necessarily tends to generate forms of concentrated economic power and severe inequality that are not consistent with these values. Liberal egalitarians conclude from this that the legitimate functions of the state exceed the more or less minimal set of functions acceptable to classical liberals. Public policies are required to limit concentrations of economic power and severe inequality. More specifically, the work of leading liberal egalitarians clearly implies a defense of technology policies beyond those proposed by classical liberals.

For liberal egalitarians the moral equality principle implies a right to participate as an equal in social and political life. Technological change should not have the social consequence of eroding this right. The "creative destruction" of technological change in capitalism threatens to do precisely that. The life prospects of the workers and communities associated with established firms and sectors can be profoundly harmed by the rise of new firms and sectors operating at (or close to) a new scientific-technical frontier. In order to minimize the social harm resulting from technological advances liberal egalitarians call for public policies supporting the transfer of new technologies into threatened communities, the extensive retraining of workers in those communities, and the provision of benefits sufficiently generous to ensure that affected individuals can continue to participate in social and political life as an equal during the transition period in which a new techno-economic paradigm is diffused.

Liberal egalitarian theorists also regard the standard classical liberal response to environmental harms associated with the use of industrial technologies as inadequate. Markets encourage concern with environmental harms when they take the form of internal costs affecting a firm's bottom line (if a rapidly depleting natural resource increases in price in a way that threatens profits, there is indeed an incentive to undertake a search for a technological substitute). A great number of environmental harms, however, are negative externalities, that is, harms that do not impact a polluting firm's bottom line. In this case market competition does not merely fail to provide an incentive to search for innovations to eliminate or minimize the harm; in so far as this search involves additional costs there is a disincentive for undertaking it. Human flourishing requires a livable environment; that too forms part of the material precondition for being able to live lives of our own choosing. If, liberal egalitarians conclude, we are to ensure that the social consequences of technological change in capitalism are normative acceptable, we must institute far more comprehensive regulation of environmentally harmful technologies than those defended in the classical liberal tradition.

A third point of contention arises from the consideration that in certain circumstances capitalist markets left to themselves will generate a rate of innovation significantly less than what is socially optimal from the standpoint of liberal egalitarianism (that is, less than what would provide the means of human flourishing to the greatest feasible extent). Technological change in a capitalist society ultimately depends upon allocations of financial capital. Classical liberals assume that the financial sector automatically functions as a means for efficiently allocating capital to the most technologically dynamic firms and sectors in the "real economy." As heterodox economists have long understood, this is not necessarily the case. Unless the financial sector is subject to effective political regulation it will tend to treat its own profits as an end in itself, generating self-sustaining speculative bubbles in the process.[2] In these circumstances increasing numbers of non-financial firms will be tempted to make profits from ownership of financial assets their ultimate end as well.[3] Past a certain point this hampers, rather than aids, technological dynamism (Perez, 2002). Funds that could have gone to investments in innovative products or processes in the "real economy" are diverted to, or remain within, the financial sector. When the speculative bubbles burst, as they always do, the production of goods and services in the "real economy" to meet human wants and needs is further harmed. Liberal egalitarians conclude that the potentially

---

[2] An inflow of funds into the market for a particular category of financial asset will raise its price. The higher price can then attract a yet higher inflow of funds into that market, raising prices yet further. The increase in (paper) wealth serves as collateral for loans, which can then also be pumped back into that market, raising prices, increasing (paper) wealth, and providing greater collateral for further loans. Human ingenuity will invariably be able to generate myriad reasons why "this time it's different," and the inflation of capital assets is not "irrational exuberance." Just as invariably, these reasons will be mistaken, and the bubble will inevitably burst. In the meantime, however, the profits that can be made from foolishly purchasing a capital asset at an overinflated price, and then selling it to a bigger fool, may greatly exceed that from alternative investments in the "real economy" (Soros, 2009).

[3] For example, retained earnings and borrowings will increasingly be devoted to stock buy-backs that raise the value of shares still outstanding (to the great benefit, we may add, of managers fortunate enough to hold stock options).

positive social consequences for human flourishing from technological change will be significantly restricted if the financial sector is not subject to effective political regulation.

Another way in which unregulated markets can distort the process of technological innovation brings us back to the notion of creative destruction. Technological change disrupts existing equilibriums. Individual firms and entire industries are threatened with being eradicated by the emergence of new firms and sectors operating close to a new scientific-technical frontier. This gives established incumbents a strong incentive to oppose the rise of new technologies, at least until they have received a minimally satisfactory return on investments in older technologies. If these incumbents possess sufficiently concentrated economic power, the rate of innovation will be significantly less than the socially optimal rate. Effective technology policies by governments can counter-act this danger. Strong antitrust legislation can prevent the stifling of new technological developments by powerful incumbents. State procurements providing a guaranteed market for innovative products can have the same effect. The use of public funds for development projects extending into to the so-called "valley of death" can also determine a rate and path of technological change superior to that which would be selected by markets alone.[4] And intellectual property rights regimes must leave ample space for newcomers, rather than act as barriers to entry protecting incumbents from innovators.

The conflict between incumbents and innovators is a major theme of Yoachai Benkler's *The Wealth of Networks*, perhaps the most important recent book examining the social consequences of technological change from a liberal egalitarian perspective. Benkler begins by noting that the technologies and forms of social organization of industrial capitalism limited the extent to which core liberal egalitarian values (autonomy, democracy, global justice) could be institutionalized.

*Autonomy* is a relatively simple matter for classical liberals; one either lacks it (slaves, indentured servants, serfs) or one does not. For liberal egalitarians, in contrast, autonomy can be a matter of degrees. When effective use of the most advanced technologies for producing goods or services requires massive investments in fixed capital, the vast majority of the populace will lack access to the requisite financial resources. They will be able to participate in the process of producing goods and services – and thereby gain access to the monetary resources necessary to support themselves and their dependents – only if those who own fixed capital grant them permission to do so. For Benkler, this need to ask permission to engage in socially productive activity counts as a significant restriction of autonomy. Further, the owners and managers of firms have an overwhelming incentive to obtain a satisfactory return on investment in fixed capital before the technologies embedded in it become obsolete, leading managers to exert control as much control as they can over workers' labor process. This too significantly restricts the degree of autonomy enjoyed by the workforce.

In the liberal egalitarian view, the moral equality principle demands that each citizen should have the opportunity to participate as an equal in public life. Open public discourse is seen as the crucial component in the *democratic will formation process* at the heart of a democratic society (Habermas, 1998). In what Benkler terms the industrial information economy, however, public discourse was profoundly restricted. The use of advanced communication

---

[4] The "valley of death" is the metaphorical place where research projects falling between basic research and research foreseen to have immediately commercializable results would otherwise languish (Wessner, 2001).

technologies required massive investment in fixed capital, and here too the vast majority of the populace lacked access to the requisite funds. Concentrated private ownership of the technological means for circulating information distorts the process of democratic will formation in two ways. First, the private owners of the means of communication have a disproportionate ability to shape public opinion, both through the content they choose to transmit (or not transmit, as the case may be) and through the manner in which this content is presented. The second difficulty is perhaps even more important. In capitalism privately owned mass media requires a mass audience to attract advertising revenue. Suppose there are three different issues a newspaper could cover, or three different programs a television station could run, the first of great interest to a large minority, the second of great interest to a different large minority, and the third of mild interest to a vast majority. The third will tend to be selected for distribution through the mass media technologies of industrialized capitalism. This counts as a systematic restriction of public discourse (Benkler, 2006, 204-11).

A third example of a tension between technological development and liberal egalitarian values concerns *global justice*. The areas of greatest market demand are not necessarily the areas of the greatest social need. Imagine two health aliments. One is a relatively minor condition that afflicts a number of affluent people in wealthy regions of the globe; the other a serious and potentially deadly affliction affecting far more people in poor regions, few of whom with significant disposable income. When pharmaceutical companies allocate investment to research proposals, priority will be given to research proposes aimed at developing drugs to address the former condition (Kremer, 2002). Similarly, research in agricultural technologies will be systematically biased in favor of innovations designed to improve the condition of affluent farmers, neglecting research on crops grown by producers in underdeveloped countries. Here too a strong case can be made that the social benefits per dollar invested would be much greater if research designed to aid those lacking purchasing power were given more priority.

Benkler is well aware that anti-trust legislation, labor regulations, publicly owned mass media, government laboratories researching medical drugs and agricultural technologies, and other familiar public policies supported by liberal egalitarians, can lessen the extent to which the technologies of industrial capitalism are associated with these negative social consequences. But he also holds that these negative consequences could not have been avoided in the past without having to sacrifice many of the benefits of liberal capitalist societies—which, all in all, have still advanced the normative values of autonomy, democratic will formation, and the fulfillment of the important social priorities better than any alternative social formation in history. Today, however, technological developments are setting the stage for the new mode of production he terms "commons-based peer production."[5] These developments have the potential to generate a quite different set of social consequences than those associated with industrial technologies.

---

[5] The contemporary significance of commons-based peer production should not be underestimated: "Ideas like free Web-based e-mail, hosting services for personal Web pages, instant messenger software, social networking sites, and well-designed search engines emerged more from individuals or small groups of people wanting to solve their own problems or try something neat than from firms realizing there were profits to be gleaned" (Zittrain, 2008, 85). Encryption software, peer-to-peer file-sharing software, sound and image editors, and many other examples can be added to this list. "Indeed, it is difficult to find software *not* initiated by amateurs" (Zittrain, 2008, 89). Individuals cooperating outside

Unlike the technologies of the industrial age, important means of production today (computers and internet connections especially) are now widely affordable, thanks to the incredibly steep upward slope of the trajectory of information technologies.[6] As a result, an ever-increasing number of individual agents are now able to own relatively advanced means of production themselves. Those who choose to develop the required expertise now have the power to decide for themselves what projects they wish to work on and with whom they wish to co-operate. By definition this expansion in the scope of free choice counts as an expansion of autonomy. Information technologies also enable a "many to many" model of communication to replace the "one to many" model of the mass media technologies of the industrial information age. This allows a tremendous expansion of the issues that can become subjects of public discourse. Democratic will-formation processes are furthered as the systematic limitations imposed by private ownership of the means of social communication are overcome. Finally, commons-based peer production greatly expands the ability of researchers with the time and expertise to collaborate effectively in developing drugs addressing the needs of those suffering from particular ailments, and seeds for farmers facing particular challenges, even if those helped by these technological products lack sufficient purchasing power to be of interest to pharmaceutical or agribusiness firms. This too clearly counts as a normative advance from the standpoint of the moral equality principle according to Benkler.

Benkler is not a technological determinist; the fact that a new mode of production is technologically possible does not imply that it will automatically emerge and grow. Commons-based peer production threatens powerful incumbents, such as media conglomerates. Incumbents can be expected to use their immense resources to push technology policies in a direction serving their interests, stifling the development of this new mode of production, thereby preventing the social consequences of contemporary technologies from being as positive as they could be. Commons-based peer production, for example, both requires free knowledge goods as inputs and produces free knowledge goods as outputs. The more knowledge goods are treated by the legal system as free public goods, the more commons-based peer production can flourish. But the more the intellectual property rights regime is extended in response to the political pressure of incumbents, the more difficult it will be for this new mode of production to mature. Liberal egalitarians, Benkler concludes, cannot be indifferent to technology policy debates. A commitment to the moral equality principle requires a political commitment to struggle against the agenda of those wishing to extend intellectual property rights in order to maintain their rents (Woo, 2010).

Liberal egalitarians are not romantics longing for a pre-modern world with little technological dynamism. They instead call for a social world in which the consequences of technological dynamism are consistent with the moral equality principle to the greatest feasible extent. In the absence of states and regimes of global governance implementing technology policies consistent with liberal egalitarian values, technological change in capitalism will necessarily tend to generate economic inequalities exceeding what is consistent with a substantive equality of civil and political liberties, a fair equality of

---

capitalist firms have also collectively produced encyclopedias that have proven useful to millions, entirely new genres of music, unprecedented access to diverse sources of information and commentary about events across the globe, and so on.

[6] The shape of this trajectory is captured in "Moore's Law," according to which computing power per dollar invested doubles every eighteen months.

opportunity, and a wide diffusion of the benefits of technological advances. To a considerable extent, then, the debate between classical liberalism and liberal egalitarianism is a debate about the social consequences of technological change in capitalism.

## 4. The Marxian analysis of technology and social change in capitalism

For Marx too there is a sense in which the individual is the fundamental unit of moral concern. His notion of "the social individual" is different from the concept of the atomistic individual found in many writings in the classical liberal tradition. But it is not so different in principle from the conception held by liberal egalitarian theorists, who echo Marx's call for "a society in which the full and free development of every individual forms the ruling principle," and in which "the free development of each is the condition for the free development of all" (Marx, 1976, 739; Marx and Engels, 1976, 506). Further, Marx does not deny that the technological dynamism of capitalism has provided the material preconditions for human flourishing to an unprecedented extent. First-time readers are often surprised by the depth of Marx's appreciation of capitalism's technological dynamism:

> The bourgeoisie, during its rule of scarce one hundred years, has created more massive and more colossal productive forces than have all preceding generations together. Subjection of Nature's forces to man, machinery, application of chemistry to industry and agriculture, steam-navigation, railways, electric telegraphs, clearing of whole continents for cultivation, canalisation of rivers, whole populations conjured out of the ground — what earlier century had even a presentiment that such productive forces slumbered in the lap of social labour? (Marx and Engels, 1976, 489).

He also agrees with the liberal egalitarian claim that in the classical liberal model the social benefits of technological dynamism are profoundly restricted. For Marx, however, the problem goes deeper than the failure to provide effective political regulation of markets. From the Marxist point of view liberal egalitarianism, no less than classical liberalism, overlooks the fundamental inversion of ends and means at the very heart of capitalism.

No one disputes that economic agents often make the acquisition of money their goal. But classical liberals and liberal egalitarians both consider money a merely proximate (short-to-medium term) end. For members of both groups money is inherently a means to make exchange more efficient and convenient, thereby serving the ultimate end of providing men and women with the goods and services they require to meet their wants and needs. The disagreement between defenders of the two viewpoints centers on whether money automatically furthers this end in forms of capitalism with only a "minimal" state and regime of global governance, or whether the more extensive state and global regime defended by liberal egalitarians is required. However, Marx asserts, in a capitalist society "use-values must never be treated as the immediate aim ... [The] aim is rather the unceasing movement of profit-making ... [t]he ceaseless augmentation of value" (Marx, 1976, 254). In a society of generalized commodity production and exchange, most units of production necessarily must make the attainment of a M' exceeding initial investment (M) their overarching goal. If they do not, over time they are increasingly likely to be pushed to the margins of social life (if not forced out of operation altogether) by units of production that do systematically make the appropriation of surplus value (the difference between M' and M) their overarching end. Most individual agents simply seek the material resources required to implement their life plans. In a society of generalized commodity exchange, however, obtaining those goods and services requires monetary resources, and this generally requires some sort of association (whether as an investor, a creditor, a wage worker, or a pensioner) with a unit of production that successfully aims at the "unceasing movement of profit-making ... [t]he ceaseless augmentation of value." Marx concludes from these considerations that the *valorization principle* is an organizing principle of social life on the level of society as a whole. Money is not a mere means adopted for the convenience of human agents; the accumulation of money capital is an end in itself.

If this is correct, then the valorization principle must also be seen as the organizing principle of technological change. Technology cannot simply be seen as a means of furthering human flourishing; in a capitalist society technology is first and foremost a means to capital accumulation. Human ends are, of course, furthered by technological change. But they necessarily tend to be furthered if and only if doing so furthers (or is at least compatible with) the end of capital accumulation.

Marx derived a set of structural tendencies regarding technological change from this starting point (Smith, 1997, 2010). The beginning of Volume 1 of *Capital* shows the need for producers to continually seek product innovations, lest their privately undertaken production turn out to have been socially wasted when competitors introduce products desired more by consumers. In Volume 2 Marx explains why advances in transportation and communication technologies necessarily tend to be sought in capitalism: they allow units of capital to complete circuits of investment, production, and sale at a faster rate, enabling more capital to be accumulated in a given period of time (Smith, 1998). Volume 3 explores how the valorization imperative is manifested in the drive to introduce innovations lowering the costs of raw materials, machinery, plants, and infrastructure. Volume 3 also sketches how technologies speeding and extending the scale of sales to consumers aid the valorization process, as do communication technologies speeding financial transactions and expanding the geographical range from which financial centers appropriate savings and to which they can transmit credit and investments.

Throughout all three Volumes Marx remarks on the threat technological development in capitalism poses to the environment. In Marx's discussion the fundamental problem stems from the discordant temporalities of capitalism and the environment of which human life is but a part. A productivity advance of, say, twenty percent, could be used to produce the same level of output in twenty percent less time. Or it could lead to the production of a twenty percent greater output in the same time. In capitalism the latter option necessarily tends to be selected, since competition among units of capital imposes the imperative to accumulate as much capital as possible as fast as possible, and this goal is generally advanced by producing and selling more commodities. Past a certain point this accelerated temporality will come into tension with the temporality of ecosystems: the capitalist economy tends to extract natural resources at a faster rate ecosystems can reproduce them, and engender wastes at a faster rate than ecosystems can absorb them. This state of affairs can continue for an extended period of time without serious difficulties arising, and technological fixes (such as the creation of substitutes for exhausted natural resources, the discovery of technical processes that use fewer natural resources, or generate less waste, or process wastes into non-harmful or even useable substances, and so on), can extend this period. Nonetheless, the underlying tension remains. For Marx, the probability that a society whose main organizing principle is "Grow or die!" (or, more exactly, "Ceaselessly augment value!") will generate environmental crises approaches one (Harvey, 1996, Part Two).

The dimension of technology and social change for which Marx is best known has to do with the labor process in the capitalist workplace. Marx thought that the pernicious effects of the valorization imperative for the satisfaction of human ends are most apparent in the capital/wage labor relation. Wage laborers are required to perform surplus labor beyond that producing an amount of value equivalent to their wages. Insofar as technologies in the workplace reduce the latter period of time, they extend the former:

> Like every other instrument for increasing the productivity of labour, machinery is intended to cheapen commodities and, by shortening the part of the working day in which the worker works for himself, to lengthen the other part, the part he gives to the capitalist for nothing. The machine is a means for producing surplus-value (Marx, 1976, 492).

Under these conditions the scientific-technological knowledge embodied in machinery is experienced by individual workers as an "alien force":

> In no respect does the machine appear as the means of labour of the individual worker ... (T)he machine, which possesses skill and power in contrast to the worker, is itself the virtuoso ... Science, which compels the inanimate members of the machinery, by means of their design, to operate purposefully as an automaton, does not exist in the worker's consciousness, but acts upon him through the machine as an alien force, as the force of the machine itself (Marx, 1987, 82-3).

Collective organization can overcome an individual worker's sense of powerlessness. But collective organization is difficult to maintain if the workforce is divided, and technological change can be used to foster such divisions: technologically-induced unemployment can set those desperate for work against those desperate to retain their jobs (Marx, 1976, Chapter 25), while communication and transportation technologies make the threat of shifting investment from one group of workers to another more effective. Technologies that deskill those enjoying relatively high levels of remuneration and control over their labor process also shift the balance in power between capital and labor in favour of the former (Marx 1976, 549). Technologies undercutting the effectiveness of strikes warrant mention as well:

> [M]achinery does not just act as a superior competitor to the worker, always on the point of making him superfluous. It is a power inimical to him ... It is the most powerful weapon for suppressing strikes, those periodic revolts of the working class against the autocracy of capital ... It would be possible to write a whole history of the inventions made since 1830 for the sole purpose of providing capital with weapons against working-class revolt (Marx, 1976, 562-3).

It is important to stress that the social consequences of technological innovation are indeterminate in particular cases. Labor history shows that the very technologies introduced to divide the work-force, deskill certain categories of workers, or break strikes, may in other contexts contribute to worker unity, enhance the skills of different workers, and help labor struggles succeed. Nonetheless, ownership and control of capital grants its holders the "operational autonomy" to initiate and direct the innovation process in the workplace (Feenberg, 2010). As long as this power is in place, Marx thought, technological change will tend to reinforce the structural coercion and exploitation at the heart of the capital/wage labor relation.

A final point regarding the workplace to be mentioned here refers again to productivity advances associated with technological change. As we saw above, the "default setting" in capitalism is for this advance to be used to increase output without sufficient regard for long-term environmental impacts. This use of technologies is also correlated with an intensification (and often extension) of the workday, despite the fact that gains in productivity could in principle be used to reduce labor time with no loss of livelihood or living standards. Marx wrote, "Since all *free time* is time for free development, the capitalist usurps the *free time* created by workers for society" (Marx, 1987, 22). Marx thought that from a world historical perspective this was the ultimately the single greatest way in which technological change in capitalism hampers human flourishing.

There is one final issue to be considered in this survey of Marx's account of the social consequences following technological change in capitalism. Marx argued that the very investments in technological change intended to further capital accumulation tend to undercut the accumulation process. New plants and firms will enter a sector when investments in fixed capital embodying more advanced technologies promise to generate above average profits due to higher levels of efficiency or products of superior quality. But as they do so, established firms do not automatically withdraw at a rate that would maintain an equilibrium of supply and demand. While the very weakest will go under, others will be content to obtain an average (or perhaps even below average) rate of profit on their circulating capital (raw materials, labor costs, etc.). There are a variety of reasons why this is a rational course for them to take. These units of production have already made the fixed capital investments (machinery, buildings, and so on) that allow profits to be won from circulating capital; if they walk away these investments will be wasted. Further, the management and work force of these firms have sector-specific skills that likely would be difficult to duplicate in any reasonable time period were the firms to shift operations to a different sector. They will also have established relationships with suppliers and distributors operating in the given sector that would be difficult and costly to establish elsewhere. They may also have relations with local governments and universities that provide important support (infrastructure, research, etc.), support they might not enjoy if they were to shift operations to a different sector. There is, finally, the hope that if they hold on they may be able at some later time to make investments in advanced technologies and leap-frog over their competitors. While these actions are rational from the standpoint of individual units of capital, they can have a collectively irrational result: the rate of profit in the sector as a whole tends to decline as the lower profits of older firms in the sector outweigh the above average profits appropriated by a relatively few newcomers (Reuten, 1991). When this dynamic occurs simultaneously in a number of key sectors, Marxists speak of an *overaccumulation crisis*. One of the most important theses of the Marxian theory of technological change is that the technological dynamism of capitalism necessarily tends to generate such crises.

The contrast between the Marxian perspective on the social consequences of technological change in capitalism and the classical liberal and liberal egalitarian viewpoints is complex. Marx actually agrees that the technological dynamism of capitalism furthers the prevision of the material preconditions for human flourishing more than any previous form of social organization. But he did not agree that technological change in an institutional context of so-called free markets and "minimal" states allows individuals the greatest opportunities of living lives of their own choosing, as classical liberals hold. Nor did he accept the liberal egalitarian assertion that all individuals could stand as an equal in social life if only the proper political regulations were in place.

No attempt will be made here to definitively resolve the controversy among these three competing perspectives. In the concluding section I shall attempt to show the continuing relevance of the Marxian account of technological change by presenting a brief historical narrative of major developments in the global economy in recent decades. My goal is to suggest that the Marxian framework illuminates important dimensions of the present historical moment overlooked in competing accounts.

## 5. Technology and social change in recent decades

Global politics today is dominated by the crisis of state finances. The public debt of many countries has reached 60% or more of their gross domestic product, and is estimated by some to soar to as much as five times GDP within a generation. In the United States and elsewhere commentators in the classical liberal tradition proclaim that their view of the predatory and profligate nature of more-than-minimal states is fully confirmed. Invoking moral obligations to future generations, they call for deep cuts in state programs, while resisting calls to raise taxes on the wealthy whose investments "create jobs and economic growth."

Writers sympathetic to liberal egalitarian values tell a different story. State deficits have metastasized primarily because of the $20 trillion of bailouts and stimulus provided by governments to the private sector in the wake of the "Great Recession" that began in 2008. The U.S. government in particular has allowed banks to exchange practically unlimited amounts of toxic assets for good money. This state spending saved the global economy from catastrophic collapse; banks have been recapitalized, and non-financial corporations have returned to profitability. But unemployment remains high, housing prices continue to decline, and the global economy remains extremely fragile. To reduce state deficits now would be to repeat the mistakes of the U.S. in the 1930's and Japan in the 90s, when weak economies were pushed back into recession by premature budget cuts. In the short term, state spending must increase to create jobs. If public investments are made in infrastructure, education, and new (especially "green") technologies, this will spur economic growth, making a reduction of state deficits in the future far less onerous than it would be today.

Liberal egalitarian analysts insist that inadequate financial regulation, the root cause of the crisis, be addressed as well. Seduced by the "efficient market hypothesis," regulators allowed paroxysms of "irrational exuberance" to generate one speculative bubble after another. The government bailouts following the bursting of these bubbles encouraged even more reckless behavior until, inevitably, the scale of the bailout overwhelmed state finances. Insufficient financial regulation also allowed debt levels in numerous regions of the global economy to exceed rational bounds, resulting in a growing and unsustainable imbalance between debtor and creditor regions. Funds lent to the former fueled consumption of imports and speculation in real estate or financial markets, neither of which generated the monetary returns necessary to repay the loans. A rebalancing of the global economy must now take place. Surplus regions must expand their domestic economies to compensate for the retrenchment of overly indebted consumers elsewhere and to help deficit regions reduce their debts through increased exports. In the worst cases (such as Greece) foreign lenders must write off many of their foolish loans (their failure to do this has brought the Eurozone project into question). Most importantly, lending and borrowing institutions throughout the globe must be regulated to ensure that such imbalances never again arise.

In the view defended here, the roots of the 2008 financial crisis can be found in the global slowdown of the 1970s after a quarter century of exceptionally high rates of investment, growth, and profits. The causes of this slowdown were varied and complex. *However, one essential element was undoubtedly the technological dynamism of Japanese and European (predominately German) producers.* In 1945 the Japanese economy was roughly a century behind the U.S., while Germany lagged a half century or so. By the 70s both had more than caught up. In many of the most technologically sophisticated sectors of the world market (consumer electronics, autos, motorcycles, chemicals, business machines, steel, and so on), these firms produced higher quality products much more efficiently than their U.S. competitors. U.S. producers in these sectors did not shut down as Japanese and German companies added to productive capacity in the global economy, and over time the rate of growth of productive capacity increased faster than the rate of growth of markets to absorb it. The result of this technological development was an overaccumulation of capital, manifested in excess productive capacity in all the leading sectors of the world market. This overaccumulation soon led to lower rates of investment, profits, and economic growth in the world market as a whole (Brenner 2006).

Economic crises are capitalism's way of renewing itself by destroying excess productive capacity through bankruptcies. The recessions of the 1970s and early 80s were certainly destructive. But there was no "Great Recession" devaluing capital investments on a scale commensurate with the problem. A number of measures allowed the global system to go down a different path, often referred to as "neoliberalism." Some of these measures essentially involved technological developments. Some did not. But they must all be considered social consequences of technological change in the sense that they were all responses to the overaccumulation difficulties brought about by technological development in the post WWII global economy.

1.  Following Nixon's unilateral decision in 1971 to in effect replace gold as the ultimate form of world money with the dollar, there was a *historically unprecedented increase in liquidity* (credit money) in the global economy in general, and the United States in particular. In principle this made it possible for markets to expand and absorb productive capacity that could not otherwise be absorbed.
2.  If increased liquidity simply set off inflation, that would not have offered the capitalist world economy a promising way forward. Accordingly, labor was disciplined through the "Volcker Shock"of 1978, a sudden rise in interest rates designed to raise the rate of unemployment in the "core" regions of the world economy beyond what had been politically acceptable previously. The pressure on real wages that followed kept inflation contained and set the stage for a significant *increase in the rate of exploitation.* The technological changes associated with the rise of "lean production" (or "flexible production") were a major part of this story (Smith, 2000). Improvements in productivity due to the introduction of information technologies into manufacturing led to waves of layoffs in the most organized sectors of the workforce. Information technologies also furthered corporate "downsizing," allowing parts of production chains to be outsourced without sacrificing management control of the production process as a whole. The advance of transportation technologies played a major role as

well in the emergence of a "networked economy" of decentralized production chains. These technological changes had a profound social consequence: the balance of power between capital and labor fundamentally shifted to the favor of the former. The increased effectiveness of the threat of unemployment from downsizing allowed capital to impose real wage cuts (despite increasing productivity), speed-ups, lengthened work days, tiered wages, and the spread of precarious employment (part time and temporary work), all of which contributed to the increase in the rate of exploitation (Basso, 2003; Head, 2003).[7] In neoliberal workplaces, no less than the factories of Marx's day, technology remained "a means for producing surplus-value."[8] Despite the contemporary rhetoric of worker "empowerment," workers' role in determining the design and use of machinery in the labor process continued to be radically restricted. The process of objectifying workers' skills in machinery accelerated with information technologies, as did the use of these technologies to continue operations during strikes. The electronic monitoring of the workforce on a massive scale is another feature of the neoliberal workplace corroborating the continued relevance of Marx's account of technological change in the capitalist workplace (Darlin, 2009).

3.  The information technologies and transportation technologies that made increased subcontracting possible also enabled production chains to extend across borders through foreign direct investment and subcontracting to locally-owned producers. The majority of foreign direct investment continued to flow from one wealthy region of the world economy to another. But Japanese foreign direct investment in China and other developing economies in East Asia exploded; more and more U.S. plants were build in Mexico, China, and elsewhere; and after the implosion of the Soviet model German capitals shifted considerable funds to investment in facilities in Eastern Europe. A relatively small portion of this production was intended for the local domestic market; most was exported back to the "core" regions of the world economy. This "globalization" of trade and investment contributed to the increase in the rate of exploitation though the increased ability of corporations to play one sector of the global labor force against another, further shifting the balance of power between labor and capital in the latter's favor (Huws, 2007; Smith, 2009). Inexpensive imports from low waged regions of the global economy offered workers in the "core" regions some compensation for the decline/stagnation of their real wages.

4.  Increased liquidity in the world market (#1) resulted in the accumulation of vast reserve of "Eurodollars" outside the U.S. and other forms of "stateless" money, as well as increased cross-border flows of financial investment. The globalization of trade and investment (#3) required companies and governments to exchange currencies on a much greater scale, while Nixon's abandoning of the gold standard led to a tripling of volatility in currency exchange markets. New financial products designed to limit the risks associated with currency fluctuations were developed. Financial firms rapidly expanded, profiting from ever-more exotic forms of financial assets. Over these decades the financial sector undertook the largest private-sector investment in information

---

[7] The anti-union policies of leading states played a key role as well.

[8] In the U.S., for example, after 1979, "The value of labor power fell for the remainder of the century (as productivity grew but hourly real wage rates for production workers did not), so that the rate of surplus value (the ratio of money surplus value to the wages of productive labor) increased by about 40%" (Mohun, 2009, 1028).

technologies, hired the greatest concentration of advanced knowledge workers, and achieved the fastest rate of product innovation in the global economy with the aid of the massive computing power they had purchased. The "financialization" of the economy exacerbated economic inequality due to the highly concentrated ownership of financial assets, further shifting the balance of power introduced in #2. The main point to insist upon is that in a world of persisting overaccumulation difficulties in non-financial sectors, financial speculation is "rational" from the standpoint of capital, and therefore must be considered as another social consequence of the technological developments that generated persisting overaccumulation difficulties.

5. The final measure defining neoliberalism is the role of consumption as an "engine" of global growth, reflected in the increasing imbalance between "deficit" regions of the world economy and "surplus regions." In the United States, wealthy households were able to go on a binge of hyperconsumerism due to the income gains they enjoyed as a result of the increase in the rate of exploitation (#2) and their gains from financial speculation (#4). Less affluent households were able to expand consumption levels despite the stagnation of real wages due to an unprecedented increase in household debt (#1) and in the appreciation of the value of their homes in the Great Housing Bubble (#4). An increasing proportion of this consumption took the form of imports, allowing exporting nations (Germany and China especially) to expand and enjoy greater trade surpluses, while the trade deficits of the U.S. and other nations (Greece, most notably) began their remorseless expansion. Banks in regions enjoying trade surpluses had a strong incentive to continue extending credit to agents in deficit countries, since that allowed domestic exporters with which they were tied to continue exporting to deficit regions. These loans also limited the appreciation of the currencies of exporting nations, which would have made their exports more expensive. With the dollar serving as world money (necessary for the purchase of oil and weapons, as well as a relatively secure store of value in an increasingly turbulent global economy) foreign investors and governments were happy to hold massive amounts of their reserves in U.S. Treasury bills (a form of credit receiving much lower returns than U.S. investors appropriated from their foreign investments). Speculative bubbles in the U.S., Ireland, Iceland, and elsewhere were fueled by foreign borrowings. The imbalances between deficit/debtor regions of the world market and surplus/credit regions increased over time to unsustainable levels, as many commentators have pointed out. But the problem goes deeper than the usual explanations, which place the primary blame either on irresponsible borrowers or on an international financial architecture that does not provide adequate supervision of cross-border financial flows. These global imbalances were a "rational" response to the need for an engine of growth in a world economy in which technological developments generated persisting overaccumulation difficulties.

From the standpoint of capital, these measures were a success. They allowed profit levels to be (partially) restored in the global economy after the slowdown of the 1970s. Levels of investment and growth in the global economy were sufficient to avoid a "Great Recession," at least in the "core" of global economy. The value of financial assets in general, and the U.S. stock market in particular, trended steeply upwards for an unprecedented period of time. The information technology revolution continued to spawn dynamic new firms and industries. The explosion of trade and foreign direct investment facilitated rates of growth in East and South Asia that were also absolutely unprecedented.

We are now, however, at a different moment in world history. The U.S. consumer market can no longer serve as the engine of growth for the global economy. With real wages stagnant, expansion of this consumer market required a constant expansion of household debt, a process that was always bound to reach a limit point. The same was true of other regions whose expanding debts provided exporting countries with growing markets.

Financialization also appears to have reached a limit point. As more and more of the profits in the global economy were appropriated by the financial sector, more and more of the credit money created there remained there (household debt, great as it was, was dwarfed by debt levels in the financial sector), and more and more of the profits of non-financial firms flowed into the financial sector (for stock buy-backers, mergers, etc.). This led to self-sustaining speculative bubbles occurring with increasing frequency and increasing scale over the course of the neoliberal period. They all eventually burst. When they did, the specter of the repressed Great Recession haunted the financial pages. But each time Central Banks came to the rescue, pumping liquidity into the financial sector and setting the stage for another round of speculative excess. The massive liquidity provided by Central Banks, and the ever-growing reserves held by surplus countries, pushed global interest rates to historical lows. Investors were happy to borrow immense sums at these rates and invest them in capital assets promising higher returns. The financial sector was happy to use its immense computing power to create ever more exotic financial products for these investors, most (in)famously by slicing and dicing "subprime" (risky) mortgages into exotic and all but incomprehensible securities ("collateralized debt obligations cubed," anyone?). Rating agencies assured investors that there was relatively little risk from purchasing these products, using computer models that assumed that the future would be like the past, despite the fact that in the past there was neither (almost) unlimited cheap credit nor financial instruments of such computer-generated incomprehensibility. The social consequences of these (mis) uses of technology were all too predictable. When a relative handful of subprime mortgages went bad, the immense edifice of global finance collapsed, revealing the fraud and collective delusion upon which it had been based. No one knew exactly which financial institutions were insolvent from their toxic loans and toxic securities; soon enough it was reasonable to think almost all were. Firms that had offered insurance against bad loans and losses in the value of securities (using the same flawed computer models) did not have anything close to the funds required to meet their obligations. The housing market collapsed, eroding the wealth of deeply indebted households to the point where additional credit was all but impossible to obtain. The specter of the long-deferred Great Recession arose once again. And once again Central Banks rode to the rescue. This time, however, the scale of the bubble was such that the scale of the bailouts threatens the solvency of those governments forced to undertake them. Here too a limit point seems to have been reached.

## 6. Conclusion

The classical liberal view fails to grasp the magnitude of the market failures that have occurred in recent decades. From a Marxian standpoint the standard liberal egalitarian position is flawed as well. Immense indebtedness, global imbalances, and recurrent financial bubbles are not accidental and irrational occurrences that could have been avoided if only regulators had fulfilled their responsibilities. Together with an increased rate of exploitation these measures were a rational (from the standpoint of capitalist rationality, at least) response to the global slowdown of the 1970s, the legacy of the technological dynamism of the post WWII "golden age." They could have been avoided only at the cost of turning the sharp but relatively brief recessions of the 1970's and early 80's into a massive destruction of excess productive capacity in the global economy.

The question now is where we go from here. If the consumer and government spending of overly-indebted economies has reached a limit point, what can take its place?

Non-financial firms sit on trillions of dollars of cash. Writers in the classical liberal tradition in the United States and elsewhere believe that radical cuts to state deficits and the eradication of "burdensome" state regulations will automatically lead to an investment binge, setting off a new period of dynamic growth in the global economy. This is pure fantasy. Overaccumulation problems have not dissipated in the global economy. In fact, new productive capacity added to the world economy in China and other developing countries has exacerbated these difficulties. Key sectors of the "networked economy" (computers, communication equipment, semi-conductors, and so on) have proven to be as susceptible to overcapacity as core sectors of the old industrial economy ever were. This explains why the rate of investment in the world market has a whole has trended downward since the 1970s, despite the recovery of profits and despite the amazing growth of investment in China and other developing countries; that growth has failed to compensate for declining or stagnating rates elsewhere (Brenner, 2006).

What of the liberal egalitarian hope that the right mix of additional government regulation and additional government stimulus could give birth to entirely new industries, dedicated to the development and use of "green" technologies? Could this provide an outlet for private investment, spurring a new "golden age" of capitalist development? There are good reasons to fear that this too is wishful thinking. This brings us to what I take to be an absolutely central issue regarding technology and social change in the contemporary global economy, the paradoxical social consequences of the spread of national innovation systems across the globe (Nelson, 1993; Smith, 2007).

Today four countries spend over three percent of their Gross Domestic Product on research and development, and another six devote over two percent of GDP of their annual economic output on R&D (*The Economist,* 2011). These nations in addition provide extensive public and private funding for scientific-technical training, public expenditures providing markets for innovations, and public policies to encourage private sector investment in advanced technologies (such as accelerated depreciation of the fixed capital that embodies technological change, a major tax break). They also possess financial sectors capable of allocating credit rapidly to start-ups operating at the technological frontier. As a result of this unparalleled proliferation of national innovation systems the moment a cluster of innovations with significant commercial potential emerges anywhere in the global economy a plethora of extensive research expenditures, tax breaks, other direct and indirect subsidies, and allocations of credit, are mobilized in a number of regions more or less simultaneously. In use-value terms this is a recipe for continued technological dynamism. In value terms, however, things are more complicated. The more national innovation systems are in place, the sooner innovating industries and sectors in the world market are threatened with overcapacity problems. This compresses the period in which high profits can be won from a competitive technological advantage. The period in which the commercialization of new innovations spurs a high rate of investment is compressed as well. In brief, a world in which effective national innovation systems have proliferated is a world of persisting overaccumulation difficulties. As long as this is the case, technological dynamism in the future (whether of "green technologies or any other sort) is not likely to lead to a new "golden age" of high rates of investment, economic growth, and real wages gains over an extended period.[9]

In this context the hope that surplus regions will expand their domestic economies to allow deficit nations to export their way out of debt is naïve. Given the facts that the ever-increasing debt card has been played, government stimulus programs cannot increase without limit, leading sectors of the world market remain plagued by serious overcapacity, and no new investment boom is likely to be inaugurated, the odds are extremely good that a considerable portion of excess productive capacity in the global economy is going to be destroyed in coming years. Political and economic elites in Germany and China, the two leading surplus/creditor nations, know that increasing the share of the world market possessed by capitals operating in their territories will lessen the odds that these capitals will be the ones devalued. Increases in domestic real wages of sufficient magnitude to compensate for the lost purchasing power in overly indebted regions would threaten this goal. Such increases would raise the prices of exports, and heighten the risk of capital fleeing Germany and China for Eastern Europe and even lower-waged areas in East Asia. Looking at the matter from another perspective, Germany and, increasingly, China have their own serious overaccumulation problems. No remotely feasible expansion of their domestic economies would be able to absorb their productive capacity; their continued growth demands the conquering of export markets (more accurately, this is a demand of capitalist rationality in the circumstances these countries find themselves in). Germany and China have reached the position they have in the world market—with Germany the unquestioned power of Europe and the Chinese economy now projected to surpass the size of the U.S.'s far sooner than anyone thought possible not long ago—due in good measure to their success in appropriating surpluses. Political and economic elites in these countries no doubt look at the U.S. and Greece and see the fate of regions that further the good of the capitalist world market by expanding domestic consumption to absorb excess productive capacity in the global economy: sooner or later they are presented a bill they cannot pay. From the standpoint of capitalist rationality, what would be rational about going down that path?

---

[9] Innovative products and processes can still be correlated with high profits for an extended period of time despite the proliferation of national innovation systems if intellectual property rights are extended in scope and enforcement. There are, however, serious social costs from doing this. It threatens to hamper innovation in society as a whole. Firms will increasingly avoid promising lines of research that might possibly infringe patents. Funds that would have otherwise gone into research will be shifted to support the armies of patent lawyers necessary to defend IPR claims and attack those of others. Fewer small firms will engage in innovative activities, lacking the funds required by the legal system. The tendency to concentrate of economic power in large corporations will be reinforced, since they are better able to fund legal costs and in a better position to come to mutually beneficial cross-licensing agreements with each other. From the liberal egalitarian perspective of Benkler, the extension of the intellectual property rights regime has another truly tragic consequence: commons-based peer production, which has the potential to become the most positive social consequence of the information technology revolution, will not be allowed to develop that world historical potential. Commons-based peer production requires that information be treated as a free public good, while the extension of intellectual property rights intensifies a commodification of information preventing this (Smith, 2012).

In these circumstances, capital's best bet for pursuing profits is a yet greater ratcheting up of the rate of exploitation[10] and household debt, and a yet more desperate search for speculative bubbles from which easy money can be made before the "smart money" gets out. From this perspective the mainstream policy debate between adherents of classical liberalism and liberal egalitarians comes down to the question whether the austerity inflicted on ordinary citizens is to be immediate and brutal, or somewhat more gradual and somewhat less brutal. Either path leads to persisting mass unemployment, a worsening gap between productivity gains and real wages, a reduction if not elimination of pensions, extended work lives, cuts to health programs, cuts to education, and cuts to anti-poverty programs at the very time rates of poverty, homelessness, and hunger skyrocket (McNally 2010).

In contemporary capitalism technological development and the productivity advances associated with it have brought about a social world in which there is *greater* material insecurity rather than less, a world where technical rationality is increasingly conjoined with social irrationality. Surely another world is possible.[11]

## 7. References

Baker, S. (2008). *The Numerati,* Houghton Mifflin Harcourt, ISBN 0-547-24793-1, New York

Basso, P. (2003). *Ancient Hours: Working Lives in the Twenty-First Century.* Verso, ISBN 978-1859845653, London

Benkler, Y. (2006). *The Wealth of Networks,* Yale University Press, ISBN 0-300-11056-1, New Haven

Brenner, R. (2006) *The Economics of Global Turbulence,* New York ISBN 978-1859847305, Verso

Darlin, D. (2009). "Software That Monitors Your Work Wherever You Are." *New York Times,* April 12 (2009).

Feenberg, A. (2010), Marxism and the Critique of Social Rationality: From Surplus Value to the Politics of Technology. In: *Cambridge Journal of Economics* Vol.34, No. 1, pp. 37-49, ISSN 0309-166X

Habermas, J. (1998). *The Inclusion of the Other,* MIT Press, ISBN 0-262-58186-8, Cambridge, MA.

Harvey, D. (1996). *Justice, Nature & the Geography of Difference,* Blackwell, ISBN 1-55786-650-5, Malden, MA

Hayek, F. (1976). *Law, Legislation, and Liberty, Volume 2: The Mirage of Social Justice,* University of Chicago Press, ISBN 0-226-320083-9, Chicago

Head, S. (2003). *The New Ruthless Economy: Work and Power in the Digital Age,* Oxford University Press, ISBN 0-195-17983-8, New York

---

[10] Baker anticipates a revolution in capital/labor relations in the near future, based on the use of massive computing power to develop mathematical models of labor processes and employees based on huge data bases. Labor processes will be disaggregated into the smallest possible fragments, and workers across the globe with the capacities to complete a particular fragment will be identified. These workers will then bid against each other for the privilege of being employed as long as it takes to complete the fragment (Baker 2008). The technologies are almost in place, Baker thinks, to extend the precariousness that already afflicts increasing numbers of the global labor force to almost all categories of workers.

[11] The main question for future research concerns the shape of an institutional framework that could combine the technological dynamism of capitalism with a higher form of social rationality. Steps in this direction have been taken in Schweickart, 2011; Smith 2000, Chapter 7, and 2009, Chapter 8.

Huws, U. (Ed.) (2007). *Defragmenting: towards a critical understanding of the new global division of labour*, Merlin ISBN 0-850-36605-4, London

Kremer, M. (2002) "Pharmaceuticals and the Developing World," *Journal of Economic Perspectives*, Vo. 16, No. 4, pp. 67-90. ISSN:0895-3309

Marx, K. (1976). *Capital, Volume I*, Penguin Books, ISBN 0-140-44568-4, New York (1987). *Economic Manuscripts of 1857-58* [the *Grundrisse*, conclusion], In: Marx, K. and Engels, F. *Collected Works: Volume 29*, International Publishers, ISBN 0-7178-0529-8, New York

Marx, K and Engels, F. (1976). The Communist Manifesto, In: Marx, K. and Engels, F. *Collected Works: Volume 6, 1845-48*, International Publishers, ISBN 978-0-7178-0506-8, New York

McNally, D. (2010). *Global Slump*, PM Press, ISBN 1-604-86332-3, Oakland

Mohun, S. (2009). Aggregate Capital Producitivty in the US Economy, 1964-2001, *Cambridge Journal of Economics* Vol. 33, No. 5, pp. 1023-46, ISSN 0309-166X

Nelson, R. (Ed.) (1993). *National Innovation Systems*, Oxford University, ISBN 0-195-07617-6, New York

Perez, C. (2002). *Technological Revolutions and Financial Capital*, Edward Elgar, ISBN 1-84064-922-4, Northhampton, MA

Rawls, J. (2001). *Justice as Fairness: A Restatement* (Second Edition), Belknap Press, ISBN 0-674-00511-2, Cambridge, MA

Reuten, G. (1991). Accumulation of Capital and the Foundation of the Tendency of the Rate of Profit to Fall, *Cambridge Journal of Economics*, Vol. 15, No. 1, pp. 79-93, ISSN 0309-166X

Schweickart, D. (2011). *After Capitalism* (Second Edition), Rowman & Littlefield, ISBN 978-0-7425-6497-8, Lanham, Maryland

Smith, T. (1997). A Critical Comparison of the Neoclassical and Marxian Theories of Technical Change, *Historical Materialism*, Vol. 1, No 1, pp. 113-33, ISSN 1465-4466 (1998). The Capital/Consumer Relation in Lean Production: The Continued Relevance of Volume II of *Capital*, In: *Essays on Marx's Second Volume of Capital*, C. Arthur and G. Reuten (Eds.), pp. 67-94, Macmillan, ISBN 0-312-21025-6, London (2000). *Technology and Capital in the Age of Lean Production: A Marxian Critique of the 'New Economy'*, State University of New York Press, ISBN 0-791-44600-X, Albany (2007). Technological Dynamism and the Normative Justification of Global Capitalism, In: *Political Economy and Global Capitalism*, R. Albritton, R. Westra, and B. Jessop (Eds.) pp. 25-42, Anthem Press, ISBN 1-84331-279-4, New York (2009). *Globalisation: A Systematic Marxian Account*, Haymarket Books, ISBN: 9781-608460236, Chicago (2010), Technological Change in Capitalism: Some Marxian Themes, In: *Cambridge Journal of Economics* Vol. 34, No.1, pp. 203-12, ISSN 0309-166X (2012). Is Socialism Relevant in the "Networked Information Age"? A Critical Assessment of *The Wealth of Networks*, In: *Taking Socialism Seriously*, R. Schmitt (Ed.), Lexington Books, Lanham, Maryland

Soros, G. (2009). *The Crash of 2008 and What it Means: The New Paradigm for Financial Markets*, PublicAffairs, ISBN 1586486993, New York

Wessner, C. (2001). The Advanced Technology Program: It Works, In: *Issues in Science and Technology*, Vol. 18, No. 1, pp. 59-64, ISSN:0748-5492

Woo, T. (2010). *The Master Switch: The Rise and Fall of Information Empires.* Knopf, ISBN 0-307-26993-0, New York

Zittrain, J. (2008). *The Future of the Internet- and How to Stop It.* Yale University Press, ISBN 0-300-15124-1, New Haven

**4**

# Reconciling Orthodox and Heterodox Approaches to Economic Growth – A Modeling Proposal

Aurora A.C. Teixeira

*CEF.UP, Faculty of Economics, University of Porto, Porto*

*INESC Porto*

*OBEGEF*

*Portugal*

## 1. Introduction

Modern growth theories have paid little attention to the understanding of the demand side phenomenon, associated to the increasing relevance of consumption activities within the society (Silva and Teixeira, 2009). In fact, the orthodox theory on economic growth neglects the aggregate demand (Dutt, 2006), whether in its shape – initially neoclassic – (Solow, 1956), or in the new or endogenous theory of growth (see Barro & Sala-i-Martin, 1995 for a survey). In the orthodox theory of economic growth, the focus lies traditionally on the logic of capital accumulation and on the influence of the different ways of technical change. However, even though technology, production and supply are essential for an economy to grow, that does not constitute the whole story. Historically, there have been, and there still are, massive changes in the products and services offered to the final demand, as well as changes in the consumers' behaviour and consumption patterns throughout the process of economic growth (Witt, 2001a).

Some authors within more heterodox tendencies, namely evolutionary (e.g., Metcalfe, 2001; Saviotti, 2001; Witt, 2001a,b), argue that what happens precisely on the demand side constitutes an essential part of the economic growth theory. According to this perspective, the consumer's role as an "innovative" being has been deeply underestimated in the Schumpeterian approaches to innovation and growth, almost exclusively based on the supply side (Metcalfe, 2001).

Trying to requalify the important role played by the demand on the process of economic growth and thus reconcile, to a certain extent, the two theoretical approaches for growth (orthodox and evolutionary) mentioned above, this chapter aims at presenting an endogenous model of growth where growth is induced by improvements in the products' quality. Using the basic models of Grossman & Helpman (1991a, 1991b), yet, on the contrary, a consumption index set up by differentiated goods and a homogeneous good, the proposed model highlights the crucial influence that the *demand* has, also granting consumers with the importance that they have in the real world.

An innovative result of the model developed for this article points out to the fact that if the weight of the differentiated goods on the consumption index is relatively lower than the market balance rate, it will always be lower than what would be socially expected (optimum), regardless of the dimension of the technological advances. Additionally, when consumers present a relatively biased consumption as far as differentiated goods are concerned, there is a possibility of situations where the market balance ratio generates excessive incentives in terms of welfare, in the event of a higher profit caused by a higher demand for differentiated goods. Therefore, these incentives will be higher than what would be socially suitable.

This article is structured as follows: in the following section, the model is briefly described. In the third section, the analytical structure of the model is developed, with the aim of determining the market balance innovation rate and the growth rate of the consumption index. We also analyse matters of welfare (Section 4) since we are in the presence of a model that involves market structures of imperfect competition (for the goods differentiated by quality, the market price is higher than the production's unit cost) and knowledge spillovers that involve scale increasing incomes. In concrete terms, the optimum growth rate is determined and, after it is compared with the market's growth rate, we draw conclusions on the insufficient or excessive incentives to R&D. Finally, the article is concluded with a summary of the model's main results. In order to facilitate reading fluency, we refer to the Annex section where we provide several technical steps for the model's result derivation.

## 2. Brief description of the model

The composition of the economic system has deeply changed throughout time (Saviotti, 2001). The observation that the system has been subject to numerous qualitative changes throughout time is an unquestionable fact for any economist (Saviotti, 1988, 1991, 1994, 1996). Modern economies include a large number of entities (products, services, production methods, competences, individual and organisational actors, institutions) that are qualitatively new and different when compared to the ones in previous economic systems (Saviotti & Mani, 1995). In fact, the economic development and growth depend on the ability that the economic system has of creating new (or new versions of) goods and services. However, for them to contribute to economic growth, such goods and services will have to be acquired by the consumers. The dynamics of the development of demand thus constitutes a fundamental aspect for economic development (Saviotti, 2001; Sonobe, Hu & Otsuka, 2004).

Using this empirical intuition, in the theoretical model described in this section, the economy grows due to the continuous improvement of a set of differentiated products, thus originating an accrual of the consumers' welfare. The utility/level of satisfaction of the consumers depends on an aggregate consumption index set up by a fixed number of differentiated products, and the increase in the level of satisfaction is exclusively determined by the quality improvements in each of those products.

The use of this economic growth approach via product innovation and based on the efforts of the companies specifically oriented to product improvement has to do with the fact that this framework is currently the one that theoretically constitutes a more complete formulation of the microeconomic basis of the aggregate phenomenon growth. Therefore, it is much more connected to the complexity of the behaviour of the economic agents that underlie the economic relations. At the same time, the use of this approach is the result (and perhaps mainly) of the empirical observation that most part of the companies' research efforts are nowadays destined to improve the products already in the market, while radical innovations (new products and new processes) are a less frequent phenomenon.[1] Following that option, the human capital factor was introduced so that its importance in a county's innovation process would be shown. As such, human capital is quite clear in the extraordinary performance of economies, such as the economies of South Korea and Taiwan, characterised by large investments in human capital (Collins, 1990).

The used structure and procedures are based essentially on the models presented by Aghion & Howitt (1992), Grossman & Helpman (1991a, 1991b) and Segerstrom (1991). Particularly, the type of explanation carried out by Grossman & Helpman (1991b) is followed closely. This is a double differentiation as far as products are concerned. There is a fixed set of differentiated products that are included in each individual's consumption basket (horizontal differentiation), and each of them is available in an unlimited number of different qualities (vertical differentiation). It is the dynamics included in the products' quality evolution that promotes economic growth.[2] The economic growth rate depends on the composition, dimension and allocation of available resources at each moment in time and, particularly, of the human capital employed in the research that produces new qualities for each differentiated product.

As we have seen in Grossman & Helpman (1991a, 1991b) and in Aghion & Howitt (1992), Research and Development (R&D) is seen as an essential activity and its success encourages the improvement in the quality of the existing products, thus promoting economic growth as well. Considering that human capital is the fundamental input of the R&D activity, it constitutes the accumulative resource that is crucial for the growth process. According to Romer (1990), human capital is the "scale variable" for economic growth. An economy with less human capital has a meaningless research sector and, therefore, it is unfit to cause improvements in the quality of products. Thus, this economy is incapable of generating economic growth.

Following Schultz (1961), Becker, Murphy & Tamura (1990), Romer (1990), Grossman & Helpman (1991b) and Barro (1991), human capital is the accumulation of effort destined to education and learning. This capital is considered to be constant at each moment, neglecting the cumulative effects that are inherent to this factor.[3] This apparently extreme assumption is justified by the fact that the model is one of technological progress. Here, the interest is focused on the relation human capital ⇒ technological progress, and not the other way around.

---

[1] According to a study from the Gabinete de Estudos e Planeamento do Ministério da Indústria e Energia (GEPIE, 1992), April 1992, based on a survey carried out on 3276 industrial undertakings (25% of the companies in mining and manufacturing industries with more than 10 employees) during the period of 1987-1989, the improvement of the existing products constituted the innovation of the most frequent product (69.1%), followed by the introduction of several new products and a new product, registering 26.8% and 15.2%, respectively. However, it is important to mention that such percentages tend to underestimate the importance of vertical differentiation since many of the new products end up replacing the ones that used to perform similar functions.

[2] Grossman and Helpman (1992, ch.3) and Romer (1990) present models that are similar to the one we are going to develop. However, in this model, economic growth is based on the increase in the number of differentiated products.

[3] Even though in a finite life horizon an individual's human capital cannot grow without a limit, the qualifications that an individual acquires may be applied to a set production technologies, from where the value of that capital will continue to increase throughout time, as well as the growth rate. This cumulative effect is neglected by assumption.

Unlike Grossman & Helpman (1991b) and Romer (1987), workforce is not seen as an input base for research activity. Even though it is plausible that an economy with a higher amount of work carries out more R&D, thus generating a higher product innovation rate, its consideration would imply that larger economies (with a higher labour/population) would, *ceteris paribus*, tend to grow more rapidly and that is precisely the result that should be avoided. Generally, a faster growth will only take place if there is an increase in the amount of factors that the economic growth promoting activities (R&D in this model) use more frequently. Thus, in this model, human capital constitutes the correct measure of an economy's scale and not the population's (Romer, 1990).

The recent theoretical literature on industrial research as an endogenous growth engine focuses on two fundamental concepts: the first concept portrays the fact that the companies that maximise profit, seek to increase their market power through the production of goods (quality goods) that are better than their direct competitors' goods. In this context, recent goods and services replace the older ones throughout time, taking advantage of a temporary profit and considering that afterwards they will also be replaced. Here, there is an implicit idea of what Aghion & Howitt (1992), following Schumpeter, designate as "the effect of creative destruction" which simultaneously promotes and limits the private value of industrial innovation; the second concept points to the fact that knowledge is a "public good" that promotes scale growing incomes for the economy as a whole. Here, the technological spillovers play a fundamental role when a company (innovator) puts a new product on the market, thus enabling researchers (potential innovators) to obtain information on production technology and on the new product's features. Thus, competitors may then start joining efforts to carry out research in order to improve the "state-of-the-art" product, even if they haven't succeeded previously while developing the product. According to Caballero & Jaffe (1993, p.16), "[i]n the process of creating new goods, inventors rely and build on the insights embodied in previous ideas; they achieve their success partly by 'standing upon the shoulders of giants'.". This way, inventions contribute to the public knowledge and that same knowledge facilitates subsequent innovation.

In this context, companies (innovators) invest in resources, hoping to discover something with a commercial value, which means that they are hoping to be capable of getting a positive profit from their research efforts. This way, in order to restore their initial investments in research, these companies must be capable of selling their products at prices that exceed the respective unit costs. This means that "... some imperfect competition in product markets is necessary to support private investments in new technologies." (Grossman & Helpman, 1994, p. 32).

The process of innovation "à la Schumpeter" represents the fact that the successful innovator (the most recent one) replaces the previous leader, taking part of the profits in the product's industry.

The modelling of the growth process on a microeconomic level shows that the process is uneven and stochastic. Companies compete amongst themselves in order to launch the new generation of the product and in certain industries there may be long periods when they do not succeed, whereas other industries may experience continuous and quick successful research. Despite that fact, on a macroeconomic level, the aggregation partly dissolves this microeconomic turbulence – and, considering the existence of a high number of quality improvements in the (fixed) set of products, the economy as a whole grows at a steady pace. Thus, research costs and benefits determine the rhythm of long-term growth.

## 3 Structure of the basic model

### 3.1 Initial remarks

Here, we will consider a continuum of differentiated products, indexed by $k$, where $k \in [0, N]$, and $N$ is constant. In order to simplify, we will normalise $N$ to 1.

Each $k$ differentiated product has its own quality and it can have an unlimited number of qualities.

Each innovation is built from the inside, which means that when a research company achieves a technological advance (product improvement) in the line of product $k$, the state-of-the-art product in that industry moves forward one generation.[4]

Where, $q_j(k) \equiv$ quality of generation $j$ of product $k$ ($j = 0, 1, 2, ...$).

Each new generation of the product provides $\lambda$ times more services than the product of the previous generation $q_j(k) = \lambda q_{j-1}(k)$ $\forall j, k$, where $\lambda > 1$ is exogenous, constant and common to all products ($\lambda \equiv$ dimension of the innovation).

The units are chosen so that the lowest quality of each product (the one available in generation 0) provides a service unit, $q_0(k) = 1$. Here, we consider that $q_j(k) = \lambda^j$ $\forall j, k, \lambda > 1$.

We define the state-of-the-art product as the most recently invented product available (i.e. the one with the highest quality), for each moment in time.

At each moment in time, entrepreneurs put their efforts on a sole product and compete amongst themselves in order to produce the next generation of that same product. Thus, the entrepreneurs that have the ability to create the state-of-the-art product or any generation prior to that product will compete amongst themselves as oligopolists. The result of this competition is a profit flow that represents the reward that the companies receive for their successful research activities. In this context, the successful research efforts promote new improvements that cause researchers to continuously compete in order to place the next generation of a certain product on the market.

As a result, the product's quality distribution goes through an upward evolution throughout time (as well as the welfare of the economic agents), and each product follows a stochastic progression of quality sequences. Despite this randomness at the level of industry, the process is constant in aggregate terms. As it will be confirmed, in the steady-state, the consumption index grows at a constant and determined rate.

Consumers, on the other hand, have at their disposal a set of differentiated goods (represented by vector $X=[x_1 \ x_2 \ ... \ x_k]$) and a homogeneous good, represented by scalar $Y$ - cash[5] -, and the price is by definition a unit price.

---

[4] Hence the expression "... 'standing upon the shoulders of giants'." by Newton and quoted by Caballero and Jaffe (1993, p. 16).

[5] The prices of $X$ have to do with the homogeneous product $Y$.

The market of the homogeneous good, $Y$, is one of perfect competition, whereas the market of the differentiated goods, $X$, is characterised by a *Bertrand* competition.

There are two production factors, unskilled labour ($L$) and human capital stock ($H$). As far as the intensity of factor use is concerned, it is assumed that the R&D activity and the production of differentiated goods, $X$, exclusively employ human capital, while the production of the homogeneous good demands human capital and unskilled labour (specific factor of $Y$).

Since the intention here is to emphasize the importance of human capital on economic growth, included in an endogenous technological progress model, we do not pay attention to important aspects such as population growth and work supply, thus avoiding questions regarding fertility, participation in workforce and variation of the working hours. We also consider that the population's human capital stock (i.e. given at any moment in time) and the human capital's fraction that is supplied to the market are exogenous. Thus, the aggregate offers of human capital stock, $H$, and labour, $L$, are fixed. Therefore, this is a model of endogenous growth in relation to technological progress and not to human capital.

## 3.2 Preferences of the consumers

The preferences of the consumers are similar to the ones proposed in Grossman & Helpman (1991b), with the additional consideration of the homogeneous good. The economy consumers provide a fixed amount of unskilled labour and human capital in exchange for a salary. At the same time, they receive interests on assets, they buy goods for consumption and save money by accumulating additional assets. While carrying out their plans, consumers take the welfare of their descendants into consideration (Barro, 1974). Therefore, the present generation maximises intertemporal utility in an infinite time frame.

Thus, the representative consumer maximises total utility, $U$, provided by the following intertemporal utility function:[6]

$$U_t = \int_0^\infty e^{-\rho t} C(t) dt$$

$$U_t = \int_0^\infty e^{-\rho t} \left[ \alpha \log u(t) + (1-\alpha) \log Y(t) \right] dt \qquad (1)$$

where

$\rho$: is the intertemporal discount rate, $\rho > 0$; [7]

---

[6] We specify a logarithmic utility function that constitutes a special case of the more general utility function, $U_t = \int_0^\infty e^{-\rho t} \frac{C(t)^{1-\sigma}-1}{1-\sigma} dt, \sigma > 0$, where the elasticity of the intertemporal substitution is unitary ($\sigma = 1$). Even though this last specification allows a wider interval of substitution elasticity, the general part of what we can take from this does not contribute much to the analysis. Therefore, we decided to sacrifice generality in order to simplify the analysis.

[7] A positive value for $\rho$ means that utility is less valued when it is received later. Ramsey (1928) assumes that $\rho = 0$, taking the maximising agent into account. The maximising agent is more of a social planner

$\alpha$: is the utility fraction of the differentiated goods on the total utility of the individual;
$u(t)$: is the aggregate consumption of the differentiated products, in moment $t$;
$y(t)$: is the consumption of the homogeneous good, in moment $t$ ;
$u(t)^{\alpha}y(t)^{1-\alpha}$: is the consumption index, in moment $t$.

The *instant utility* for each individual on the aggregate consumption of $X$ is provided by:

$$\alpha \log u(t) = \alpha \int_0^1 \log\left(\sum_j q_j(k)d_{jt}(k)\right) dk \; , \tag{2}$$

where

$q_j(k)$: is the quality of generation $j$ of product $k$;
$d_{jt}(k)$: is the consumption of product $k$, by generation $j$, in moment $t$ .

The sum in (2) includes the set of generations of product $k$ that are available in moment $t$. It is important to highlight that quality, $q_j$, positively contributes to utility.

Consumers are subject to an *intertemporal budgetary constraint* provided by:

$$\int_0^\infty e^{-R(t)}[E(t) + Y(t)]dt \le A(0) \tag{3}$$

where

$E(t)+Y(t)$: is the total expenditure, in moment $t$ ;
$R(t)$: is the factor of the accumulated interest until moment $t$, $R(t) = \int_0^t r(\tau)d\tau$ ;
$A(0)$: is the current value of the factors' income flow, plus the assets initially held by the individual.

On the other hand, $E(t)$ is the total expense in differentiated products and $Y(t)$ is the expense with the homogeneous good, respectively provided by

$$E(t) = \int_0^1 \left(\sum_j p_{jt}(k)d_{jt}(k)\right) dk \tag{4}$$

$$Y(t) = y(t)p_Y(t) = y(t) \; considering \; that \; p_Y(t) = 1 \tag{5}$$

where

$p_{jt}(k)$: is the price of a unit of product $k$ from generation $j$, in moment $t$;
$d_{jt}(k)$: is the consumption of product $k$ by generation $j$, in moment $t$;
$p_Y(t)$: is the price of the homogeneous good, $Y$ (cash), in moment $t$.

The goal of the representative consumer is to maximise (1), respecting the budgetary constraint (3). Considering that we are dealing with regular goods with unitary elasticity

than a consumer that chooses to consume and save, whether from the current generation or from future generations.

income, the consumer's optimisation process may be (by analytical convenience and easiness of exposure) divided into three stages:

1. Given the expense aggregated in differentiated goods at a certain moment in time, $E(t)$, it is necessary to find the optimum allocation of this expense among those goods;
2. Given the total expense at a certain moment in time, $E(t)+Y(t)$, it is necessary to find the optimum distribution between $E(t)$ and $Y(t)$;
3. Taking into consideration the allocation of the expense that maximises the consumer's utility at each moment (whether it is on the differentiated products' shares in $E(t)$, or on the $E(t)$ and $Y(t)$ shares in the total expense), it is necessary to find the distribution of $E(t)$ throughout time in order to maximise the intertemporal preferences.

As far as the *static allocation of the aggregate expense* for the differentiated goods is concerned, the consumer maximises instant utility, selecting a sole generation $j$ for each of those goods with the lowest price adjusted by quality. This means that $J_t(k)$ is selected, such that $p_{jt}/q_{jt} = min\{ p_{jt}(k)/q_j(k); j = 0, 1, ... \infty \}$.

Such procedure will originate the *static demand functions* for differentiated products,

$$ d_{jt}(k) = \begin{cases} \dfrac{E(t)}{p_{jt}(k)} & , \quad para \quad j = J_t(k) \\[3mm] 0 & , \quad para \quad j \neq J_t(k) \end{cases} \tag{6} $$

As a consequence, the *instant utility* for the differentiated products is obtained by

$$ \alpha \log u(t) = \alpha \int_0^1 \left( \log \sum_j q_j(k) \frac{E(t)}{p_{jt}(k)} \right) dk \text{ , or by} $$

$$ \alpha \log u(t) = \alpha \int_0^1 \left( \log \sum_j \frac{q_j(k)}{p_{jt}(k)} + \log E(t) \right) dk \text{ .} $$

Since the consumer selects only the leader quality for each product (and supposing that the price is uniform), (i.e. the *state-of-the-art* product), then:

$$ \alpha \log u(t) = \alpha \log E(t) + \alpha \int_0^1 \left( \log \frac{q_t(k)}{p_t(k)} \right) dk \tag{7} $$

$$ \alpha \log u(t) = \alpha \log E(t) + \alpha \int_0^1 (\log q_t(k) - \log p_t(k)) dk \text{ .} $$

In order to maximise its global instant utility (including the differentiated products and the consumption of the homogeneous good), the consumer distributes the total expense $(E+Y)$ proportionately between differentiated goods and the homogeneous good (see Annex):

$$\frac{\alpha}{1-\alpha} = \frac{E(t)}{Y(t)} \tag{8}$$

This way, the expressions of utility and budgetary constraints of the representative consumer are obtained as follows:

$$U_t = \int_0^\infty e^{-\rho t}\left[ \log E(t) + \alpha \int_0^1 (\log q_t(k) - \log p_t(k))\,dk + (1-\alpha)\log\left(\frac{1-\alpha}{\alpha}\right)\right]dt \tag{1'}$$

$$\frac{1}{\alpha}\int_0^\infty e^{-R(t)}E(t)\,dt \le A(0). \tag{3'}$$

The solution for the consumer's *dynamic problem* consists of determining the maximum intertemporal utility function (1'), which is subject to the budgetary constraints (3'),

$$\underset{\{E(t)\}}{Max\,U_t} \quad s.a. \quad \frac{1}{\alpha}\int_0^\infty e^{-R(t)}E(t)\,dt \le A(0).$$ The solution is the following optimum temporal

trajectory of the expenses (see Annex):

$$\frac{\dot{E}}{E} = r - \rho, \tag{9}$$

where $r$ is the interest rate, at a moment in time and $\dot{E} = \dfrac{dE}{dt}$.[8]

### 3.3 Supply-side – The producers

As it was mentioned in the beginning of this section, there are two production factors in this economy: unskilled labour, $L$, and human capital, $H$. It is also admitted that, given the strong emphasis on human capital, there is a particular case related to the frequency of use of the factors for each of the considered sectors: the R&D and differentiated product sectors use human capital on an exclusive basis, while the sector of the homogeneous good uses human capital and unskilled labour (this input is specific for this sector).

For the activities of research and production of differentiated goods, we admit fixed coefficient technologies. Thus, in order to carry out research activities, an R&D unit requires $a_{HI}$ human capital units, while a unit that produces differentiated good requires $a_{HX}$ human capital units. As far as the production of the differentiated goods is concerned, $Y$, we consider a neoclassical technology with scale constant incomes and substitutability between the inputs ($L$, $H$) that are necessary for the production.

### 3.3.1 Industry of the homogeneous good

Since the market structure for the homogeneous good, $Y$, is one of perfect competition and since all of the unskilled work is used in its production, the producers, profit maximisers,

---

[8] An identical solution will henceforth be used in order to reference a derivative of a variable in time ($t$).

produce an amount of $Y$, such that $p_Y = c_Y(w_H, L)$ , where $c_Y(.)$: is the marginal cost of the production of good $Y$; $p_Y$: is the price of good $Y$ (provided); $w_H$: is the price of each human capital unit. Since good $Y$ is cash, $1 = c_Y(w_H, L)$.

### 3.3.2 Industry of differentiated products

At this point, we followed the balance derivation carried out by Grossman & Helpman (1991a, 1991b) assuming that all companies compete as far as prices are concerned. There are laws on patents that indefinitely protect the intellectual property rights of the innovators (which provide the innovative companies the exclusive right to sell the goods that they invent), so patent licensing is not possible. This way, it is guaranteed that the whole production of differentiated goods is carried out by companies that have successfully developed new state-of-the-art products.

In order to describe the price-fixing process in this industry, we will imagine that there are two different companies: one (leader) has access to state-of-the-art technology, while the other one (follower) is capable of producing the good that is inferior to the leader product in quality terms.

*1st Situation*

At a given moment, the follower company fixes the price, $c_X$, which is the lowest price consistent with non-negative profits. The consumers are willing to pay a premium for a state-of-the-art product. However, they will choose the product of the previous generation if the leader fixes a price that exceeds $\lambda c_X$ ($\lambda$: accrual in services /quality provided by the leader good relatively to the closest rival product).

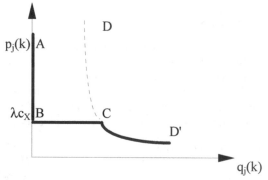

*Source*: Grossman and Helpman (1993, p. 90)

Chart 1. Curve of the *demand perceived* by the industry leader

When the leader fixes a price lower than $\lambda c_X$, the whole industry demand is satisfied on its own since its product offers a price (per quality unit) that is lower than the follower's price. If the prices fixed by the leader and by the follower are $\lambda c_X$ and $c_X$ respectively, then the leader can sell any amount throughout the BC segment. This way, the leader's optimum response should be to establish a price that is infinitesimally lower than $\lambda c_X$ in view of the $c_X$ price fixed by its closest competitor. There are three reasons for that: 1) with that "limit price", it is possible to keep the other company away from the market; 2) if they set a price higher than $\lambda c_X$, they will lose clients to the follower; 3) if a price slightly lower than $\lambda c_X$ is set, that is not a rational decision because the functions of industry demand have a unitary elasticity, and it does not allow gains in the marginal revenue. With this behaviour, the leader absorbs the whole industry demand and thus it will be possible for the price to become closer to the monopoly price that the competition allows.[9]

## 2nd Situation

When the leader fixes the $\lambda c_X$ price, the optimum response of the follower will consist of fixing the $c_X$ price by two orders of reason: 1) if the price decreases (if it is set below the marginal cost), there will be losses; 2) if the price increases, the follower will be at a position of indifference, considering that the sales will be inexistent.

Therefore, as we can see in Grossman & Helpman (1993), the balance of Bertrand-Nash is reached when the leader practices the "limit price", thus eliminating its most direct competitor (follower) from the market.

Since this balance may be extended to the most general case where the technological leaderships may overcome a generation ($\lambda^i$), we can conclude that the state-of-the-art product always has a lower (quality adjusted) price.

Considering what has been said regarding the technology for the development of the product and the nature of the property rights, it is guaranteed that in each industry and at each moment, there is only one leader. The leader is always one step ahead of its closer competitor (whichever the industry).

From what has been previously said, we can conclude that all state-of-the-art products (leaders in quality) for each industry have the same limit price [ $p(k) = p$ ]:[10]

$$p = \lambda c_X , \tag{10}$$

where

$p$: is the row vector ($1 \times k$) of similar prices for the differentiated products $k$;
$c_X$: is the row vector ($1 \times k$) of similar marginal costs for the differentiated products $k$.

Replacing (10) with (6), we obtain the following demand functions:

$$d_{jt} = \begin{cases} \dfrac{E(t)}{p} = \dfrac{E(t)}{\lambda c_X}, & se \quad j = J_t(k) \\[2em] 0, & se \quad j \neq J_t(k) \end{cases} \tag{11}$$

The monopoly profit flow is obtained by:

---

[9] Barro & Sala-i-Martin (1995, ch. 7) present a detailed analysis of the differences between the limit price and the monopoly price, in the context of this type of model.
[10] This refers to differentiated products with a similar marginal cost.

$$\pi = pX - c_X X = pX - \frac{p}{\lambda}X = \left(1 - \frac{1}{\lambda}\right)pX = \left(1 - \frac{1}{\lambda}\right)E \ . \tag{12}$$

In balance, leaders do not invest any resources in order to improve their own state-of-the-art products. Therefore, the innovations are carried out by the followers. When the followers succeed, they will be one step ahead of the previous leader.[11]

However, it is important to highlight that R&D is a risky activity. A company may put all its research efforts on any state-of-the-art product. If this company uses its resources to invest in R&D with an intensity of $i$ during a time interval that lasts $dt$, there is a $idt$ probability of success for the development of the product's next generation. Thus, the success in the R&D activity involves a *Poisson* probability distribution, with a rate of occurrence that depends on the level of the R&D activity. In order to reach a $i$ R&D intensity, the company must invest $a_{HI}i$ human capital units per time unit. This way, the probability that the company has of succeeding in its research is strictly proportional to the invested resources (human capital).

Analysing the state-of-the-art products available in the market, the potential producers may obtain precious technical information that allow them to fulfil their own research efforts.[12] When the potential producer (follower) succeeds in its research, the company will take the leadership of the industry of the selected product, thus taking advantage of a profit flow provided by (12). This profit will cease as soon as the next research success case is achieved (by other company) for the same product line.

The profit flow in equation (12) is identical for all industries $k$. Thus, as long as the expected duration of the leadership is identical as well, companies are indifferent as far as the industry in which they apply their research efforts is concerned. Like Grossman & Helpman (1991b, 1993), we consider that there is a symmetrical balance where all the products are subject to the same aggregate R&D intensity ($i$), and that is why that indifference occurs.[13]

*Free entry condition*

We will consider that $v$ is the updated value of the profit (uncertain) that will flow to the industry's leader, i.e. the market value of the leader company or, similarly, the amount of the premium guaranteed by the success in R&D.

A company may obtain $v$ with a probability of $idt$ if it invests an amount of $a_{HI}i$ resources during a time interval of $dt$, incurring in a cost of $w_H a_{HI} idt$.

---

[11] Leaders prefer to earmark the resources for the development of a leadership position in other markets instead of extending their advance (in terms of quality) in the market where they are already included. The reason for this is that the accrual in profit that would stem from this situation - $\Delta\pi = \left(1 - \frac{1}{\lambda^2}\right)E - \left(1 - \frac{1}{\lambda}\right)E = \left(1 - \frac{1}{\lambda}\right)\frac{E}{\lambda}$ - is strictly lower than the accrual in profit that non-leaders would obtain if they were to succeed in their research efforts, that is, $\left(1 - \frac{1}{\lambda}\right)E$ .

[12] This reflects the abovementioned spillover benefits of innovation, which are related to the nature of public good for technology.

[13] Since researchers hope that future innovations have an equal probability in the different industries (i.e. that the flow of profits has an equal duration in each industry), they will be indifferent to the choice of industry.

The company's financing is carried out through the issuing of securities, which grant the holders with the income flow associated to the industry's leader if the research effort is successful. If, on the contrary, the research effort fails, the security holders incur in a loss that is equivalent to the total of the invested capital, $v$. However, the security holders may neutralise the risk by holding diversified bases. For that, each of the companies must maximise the net profit expected for the respective research efforts, $vidt - w_{H}a_{H}idt$.

Therefore, the new company should choose its research intensity in order to maximise the net benefit expected from that same research. This way, whenever the current value of the profit, $v$, is lower than the amount invested in research, $a_{HI}w_{H}$, there is no investment ($i = 0$). On the other hand, this investment tends to infinite when $v > a_{HI}w_{H}$.

With a positive, yet finite balance of research investment, we verified that $v = a_{HI}w_{H}$, a case where research companies are indifferent to the scale of their research efforts. This way, the condition for free entry in the R&D sector is

$$a_{HI}w_{H} \geq v \text{ with uniformity every time that } i > 0 \tag{13}$$

The industry leaders generate a flow of dividends of $\pi dt$ during a time interval that lasts $dt$. If none of the competitor research efforts succeed during that period of time, the shareholders of the leader company will benefit from capital earnings to the amount of $\dot{v}dt$. This means that the shareholders will obtain $\dot{v}dt$ capital earnings with a probability $(1-idt)$, admitting that the companies' research efforts are statistically independent. However, the leader product may be improved during the $dt$ interval, with a probability of $idt$. In this last case, the shareholders will suffer losses in the amount of the total invested capital, $v$. This way, the expected rate of return for the company's shares per time unit is $\dfrac{\pi + \dot{v}}{v} - i$.

Since the research results in the different industries are not contemporarily correlated, the risks that the leader faces are idiosyncratic. Therefore, shareholders may obtain a certain income if they have a diversified share base for each company in the different industries.[14]

Thus, the market tends to value the companies where the rate of return of the shares is exactly equivalent to the interest rate on the obligations without risk, $r$.

$$\frac{\pi + \dot{v}}{v} - i = r \tag{14}$$

The previous equation reflects the non-arbitrage condition.

### 3.4 Balance of the goods and labour markets

As it has been mentioned before, a research unit requires $a_{HI}$ human capital units. Thus, the total employment of this factor in the research activity equals $a_{HI}i$.

In the production of differentiated goods, each output unit requires $a_{HX}$ human capital units. In order to simplify, yet without sacrificing generality, we assume that $a_{HX} = 1$. The

---

[14] Here, there is no statistical correlation between the possibilities of success of the several contemporary research efforts, which causes no harm to the existence of spillovers.

aggregate demand of human capital carried out by the producers of these goods is represented by $\dfrac{E(t)}{\lambda c_X}$ , since each industry of differentiated products produces $\dfrac{E(t)}{p} = \dfrac{E(t)}{\lambda c_X}$

[see (11)], and since the sector's dimension is standardised by construction ($k \in [0,1]$). Because the production of these goods uses exclusively human capital as an input, $c_X = w_H$.

In order to produce a product unit, the industry of the homogeneous good, $Y$, requires human capital, $H$, and unskilled labour, $L$. *Shephard's lemma* states that this industry's demand for human capital and unskilled labour should be $a_{HY}(w_H, w_L)$ and $a_{LY}(w_H, w_L)$, respectively, per product unit, where $a_{jY}(.)$ is the requirement of the factor $j$ ($j = H, L$) in the production of $Y$. This is equal to the partial derivative of $c_Y(.)$ in order of $w_j$.

Generically, the balance conditions for the factor market are $a_{HI}i + a_{HX}X + a_{HY}(w_H, w_L)Y = H$ and $a_{LY}(w_H, w_L)Y = L$ . In the actual case,

$$a_{HI}i + \frac{E(t)}{\lambda w_H} + H_Y(w_H, L) = H$$

$$\tag{15}$$

$$a_{LY}(w_H, w_L)Y(w_H, L) = L$$

where

$Y(w_H, L)$: is the supply of the homogeneous good, $Y$;

$H_Y(w_H, L)$: is the demand of human capital for the production of the good, $Y$, when the total unskilled labour is employed.

On the other hand, from Expense = Income, we can extract the following balance condition for the market of goods (differentiated and homogeneous):

$$E(t) + Y(t) = w_L L + w_H H + \pi(t) - a_{HI} w_H i(t). \tag{16}$$

This condition, combined with condition (8) for the maximisation of instant utility, $\dfrac{\alpha}{1-\alpha} = \dfrac{E(t)}{Y(t)}$ , and the expression (12) of the monopoly profit flow $\pi(t) = \left(1 - \dfrac{1}{\lambda}\right)E(t)$ , originates

$$E(t) = \frac{w_L L + w_H H - a_{HI} w_H i(t)}{\left(\dfrac{1}{\alpha} + \dfrac{1}{\lambda} - 1\right)} , \tag{17}$$

i.e., the (simplified) balance condition in the market of goods (differentiated and homogeneous).

## 3.5 Market balance innovation rate

Taking into consideration the equations that describe the condition for consumption optimisation (9), the expression of profit (12) and the non-arbitrage condition (14), we have solved the system in a way that would make it possible to find the intensity of the balance research effort (i.e. the innovation rate). Thus, replacing (9), which represents the optimum time line of the expenses $\frac{\dot{E}}{E} = r - \rho$, and (12), which represents the monopoly profit flow $\pi(t) = \left(1 - \frac{1}{\lambda}\right) E\,(t)$, in the condition of non-arbitrage (14), $\frac{\pi + \dot{v}}{v} - i = r$, the result is as follows:

$$\frac{\dot{v}}{v} = \rho + i + \frac{\dot{E}}{E} - \frac{\left(1 - \frac{1}{\lambda}\right)E}{v}. \tag{18}$$

In the steady-state balance, $\frac{\dot{v}}{v} = \frac{\dot{E}}{E} = 0$, what was replaced in (18) and in (9) implies, respectively:

$$\left(1 - \frac{1}{\lambda}\right)E = (\rho + i)v \tag{19}$$

$$r = \rho. \tag{20}$$

From equation (19), we conclude that $E = \dfrac{(\rho + i)v}{\left(1 - \dfrac{1}{\lambda}\right)}$ , combined with the free entry condition

(13), $a_{HI}w_H = v$, originates $E = \dfrac{(\rho + i)a_{HI}w_H}{\left(1 - \dfrac{1}{\lambda}\right)}$ . By combining this last equation with the

balance condition of the market of goods, (17), $E(t) = \dfrac{w_L L + w_H H - a_{HI}w_H i(t)}{\left(\dfrac{1}{\alpha} + \dfrac{1}{\lambda} - 1\right)}$ , after some

algebraic manipulation, we obtain the market's balance innovation rate:

$$i^e = \frac{\alpha\left(1 - \dfrac{1}{\lambda}\right)\left(\dfrac{w_L}{w_H}L + H\right)}{a_{HI}} - \frac{\rho}{\lambda}[\lambda - \alpha(\lambda - 1)]. \tag{21}$$

Give the specific nature of the use of the unskilled labour factor, this balance innovation rate is the result of the entrepreneurs' decentralised choice that consists of using (in the most lucrative way) the human capital that is not used in the production of the homogeneous goods for the production of differentiated products *versus* production of innovations (research activities).

As we analyse in Section 3.7, the market balance innovation rate is all the higher as the economy's investment in human capital increases ($H$); on the other hand, it will be all the lower as the economy's investment in unskilled labour decreases ($L$), admitting that the substitution elasticity is lower in the production of the homogeneous goods; all the higher as the research productivity $(1/a_{HI})$ is higher; as the families are more patient (and as the intertemporal preference rate is lower); as the dimension of technological advances is higher ( quality "ladder"); and as the consumers' degree of "sophistication" is higher $(\alpha)$.

### 3.6 Growth rate of the consumption index

Taking into consideration the equations referring to the instant utility of the aggregate consumption of $X$ (2), the static demand functions for the differentiated products, (6), the balance of the market of goods (8) and the limit price fixing (10), it is possible to obtain (note that the expression on the left is the logarithm of the consumption index):

$$\alpha \log u(t) + (1-\alpha)\log y(t) = \alpha \int_0^1 \left( \log \sum_j q_j(k)dk \right) + \log E(t) - \alpha \log \lambda - \alpha \log w_H +$$

$$(22)$$

$$+(1-\alpha)\log\left(\frac{1-\alpha}{\alpha}\right)$$

Let us now consider that

$f(j,t)$: is the probability that any product $k$ has of registering $j$ quality improvements, during the time interval $t$.

Since, in balance, every product has the same research intensity, $f(j,t)$ represents the fraction of products that are improved $j$ times before interval $t$. This means that the counting of all products and possible generations will generate (Grossman & Helpman, 1991b)

$$\int_0^1 \log q_t(k)dk = \sum_{j=0}^\infty f(j,t)\log \lambda^{\ j} .$$

$$(23)$$

As $\sum_{j=0}^\infty f(j,t)\log \lambda^j = \sum_{j=0}^\infty f(j,t)j\log \lambda = E(j)$, by the properties of *Poison's* distribution, it generates[15]

$$\int_0^1 \log q_t(k)dk = it\log \lambda .$$

$$(24)$$

Replacing (24) with (22) we obtain:

$$\alpha \log u(t) + (1-\alpha)\log y(t) = \alpha it\log \lambda + \log E - \alpha \log w_H + (1-\alpha)\log\left(\frac{1-\alpha}{\alpha}\right) - \alpha \log \lambda \qquad (25)$$

---

[15] By Poisson's distribution properties, $f(j,t) = p(j) = \dfrac{(it)^j e^{-it}}{j!}$ and E(j) = $it$, where $E$ represents here the expected value (see Santos, 1988, p. 245).

Thus, the growth rate of the consumption index is (note that, in balance, $E$ and $i$ are constant, while $w_H$ is globally determined by the provision of factors that we consider to be fixed):

$$g \equiv \frac{d\log\left[u(t)^{\alpha} y(t)^{1-\alpha}\right]}{dt} = \alpha i \log \lambda . \tag{26}$$

## 3.7 Compared static analysis of the market balance

It is important to remember that, in (21), $i$ is provided by

$$i^e = \frac{\alpha\left(1-\dfrac{1}{\lambda}\right)\left(\dfrac{w_L}{w_H}L+H\right)}{a_{HI}} - \frac{\rho}{\lambda}[\lambda-\alpha(\lambda-1)] .$$

Thus,

$$g = \alpha\log\lambda\left\{\frac{\alpha\left(1-\dfrac{1}{\lambda}\right)\left(\dfrac{w_L}{w_H}L+H\right)}{a_{HI}} - \frac{\rho}{\lambda}[\lambda-\alpha(\lambda-1)]\right\} . \tag{27}$$

From the analysis of (26), we conclude that the economy's growth rate ($g \equiv$ growth rate of the consumption index) is all the higher

a.    as the innovation rate $\dfrac{\partial g}{\partial i} = \alpha\log\lambda > 0$ is faster.

b.    as $\alpha$, which is the weight of the differentiated products in the consumption index, is

higher, $\dfrac{\partial g}{\partial \alpha} = i\log\lambda + \dfrac{\partial g}{\partial i} \cdot \dfrac{\partial i}{\partial \alpha} = i\log\lambda + \alpha\log\lambda\left[\dfrac{\left(1-\dfrac{1}{\lambda}\right)\left(\dfrac{w_L}{w_H}L+H\right)}{a_{HI}} + \dfrac{\rho}{\lambda}(\lambda-1)\right] > 0$.

This means that the more sophisticated the consumers are, in a sense that they have a larger preference for quality differentiated products (i.e. they are more sensitive to quality), the higher is the economy's growth rate.[16]

From the analysis of (27), we additionally conclude that the innovation rate (therefore, the growth rate) is all the higher

c.    as the economy's investment in human capital is higher

---

[16] It would be interesting to consider the chance that the highest sophistication of the consumers is positively correlated to the level of human capital. In that case (we do not analyse that case in particular here), the human capital would also have an influence on demand since $\alpha$ would be a positive function of $H$. Therefore, we would be in the presence of an additional influence channel for human capital on economic growth.

$$\frac{\partial i}{\partial H} = \frac{\alpha\left(1-\frac{1}{\lambda}\right)}{a_{HI}}\left[1+\frac{\left(\frac{\partial w_L}{\partial H}w_H - \frac{\partial w_H}{\partial H}w_L\right)L}{\left(w_H\right)^2}\right] > 0 \quad \text{since} \quad \frac{\partial w_L}{\partial H} > 0 \text{ e } \frac{\partial w_H}{\partial H} < 0.$$

The remunerations of the factors are fixed for each level of factor provision. Thus, an increase in the amount of available human capital will cause a relative shortage of $L$, thus leading to an increase in this factor's remuneration. On the other hand, given the relative abundance of human capital, this factor's remuneration tends to decrease, thus reducing the costs of research activities and increasing the growth rate.

d.    as the economy's investment in unskilled labour is lower, admitting that the substitution elasticity in the production of the homogeneous good is not too high.

$$\frac{\partial i}{\partial L} = \frac{\alpha\left(1-\frac{1}{\lambda}\right)}{a_{HI}}\left[\frac{w_L}{w_H} + \frac{\left(\frac{\partial w_L}{\partial L}w_H - \frac{\partial w_H}{\partial L}w_L\right)L}{\left(w_H\right)^2}\right] < 0 \quad \text{since} \quad \frac{\partial w_L}{\partial L} < 0 \text{ e } \frac{\partial w_H}{\partial L} > 0$$

$$\text{and admitting that } \left|\frac{\partial w_L}{\partial L} - \frac{\partial w_H}{\partial L}\frac{w_L}{w_H}\right| > \frac{w_L}{L}$$

This result comes from the fact that the homogeneous good, $Y$, requires human capital ($H$) and unskilled labour ($L$) in its production. Therefore, an increase in this sector's specific factor – the $L$ factor – decreases the respective remuneration and increases the remuneration of the human capital since it becomes relatively scarce. Considering that the effect of the increase in the relative remuneration of $H$ is higher than the effect of the replacement of $H$ by $L$ in the production of $Y$,[17] the innovation rate will fall, as well as the economy's growth rate.

e.    as the research productivity ($1/a_{HI}$) is higher, $\dfrac{\partial i}{\partial a_{HI}} = -\dfrac{\alpha\left(1-\frac{1}{\lambda}\right)\left(\frac{w_L}{w_H}L+H\right)}{\left(a_{HI}\right)^2} < 0$.

An increase in productivity increments the expected return of the research activity, thus encouraging efforts/investments in R&D and, consequently, by increasing the innovation rate, it increases growth.

f.    as the families are more patient (when the intertemporal preference rate is lower, $\rho$)

$$\frac{\partial i}{\partial \rho} = -\frac{1}{\lambda}[\lambda - \alpha(\lambda-1)] < 0.$$

---

[17] Globally, the increase of $L$ makes the human capital relatively scarce. However, at the same time, it releases part of the human capital for the production of differentiated goods and for research. What we admit here is that the substitution elasticity for the production of $Y$ is not high enough to make the second effect dominant.

When families adopt a behaviour that is characterised by savings, it contributes to an increase in the innovation rate and, consequently, the growth rate.

Finally, from (25) and (26), it was possible to extract the effect of an increase in the dimension of the quality technological advances ($\lambda$). Growth is all the higher:

g.  as the dimension of the technological advances is higher (quality "ladder")

$$\frac{\partial g}{\partial \lambda} = \frac{\alpha i}{\lambda} + \alpha \log \lambda \frac{\partial i}{\partial \lambda} = \frac{\alpha i}{\lambda} + \frac{\alpha^2 \log \lambda}{\lambda^2}\left[\frac{\left(\frac{w_L}{w_H}L + H\right)}{a_{HI}} + \rho\right] > 0 .$$

It is important to remember that an increase in the dimension of the quality advances ($\lambda$) promotes growth in two ways: directly, since the quality ladder is higher; or indirectly, which leads to a faster occurrence of technological advances, therefore increasing the expected return of the research. This will attract resources (human capital) to R&D activities.

h.  as the consumers' degree of "sophistication" is higher ($\alpha$)

$$\frac{\partial i}{\partial \alpha} = \left[\frac{\left(1 - \frac{1}{\lambda}\right)\left(\frac{w_L}{w_H}L + H\right)}{a_{HI}} + \frac{\rho}{\lambda}(\lambda - 1)\right] > 0 .$$

This analysis has an important consequence, which is the prevision (*ceteris paribus*) of a growth rate that is lower for the larger economies (those with a larger amount of unskilled labour). On the other hand, the dimension of the human capital is crucial. The higher this dimension is, the faster will growth be.

This is in agreement with Romer's conclusions (1990) "... that the correct measure of scale is not population but human capital..." (p. S78) and that an economy "... with a larger total stock of human capital will experience faster growth." (p. S99).

In that context, a large economy with a large amount of skilled labour ($H$), carries out more industrial research because the R&D sector uses this factor more frequently. Such economy will grow faster than the other economy that has similar features, yet a smaller amount of human capital. However, a large economy, largely set up by unskilled individuals (high $L$), may grow more slowly than the other, which is identical, yet with a smaller population.

In other words, Grossman & Helpman (1994, p. 36) mention the same conclusion:[18] "[t]he larger labor-abundant country, which specializes relatively in labor-intensive

---

[18] These conclusions implicitly suggest that, as it was mentioned above in the revision of the literature, the international free trade could contribute to accelerate growth. For a more complete and detailed analysis on the accumulation of human capital and the interconnections of human capital with the matters of international trade, see chapter 5 of Grossman & Helpman (1993).

production, might will conduct absolutely less industrial research than a smaller country with comparative advantage in R&D.".

## 4. Conclusion

Trying to requalify the key role played by the demand on the process of economic growth, this article is the algebraic result of an endogenous growth model where growth is induced by improvements in the products' quality.

Similarly to what happened in the models of Grossman & Helpman (1991a, 1991b), each innovation is built from the previous one and the same thing happens to each generation. Each new generation of the product proportionately supplies more services than the previous generation (i.e., we admit that, in a reasonable way, the dimension of innovation, identified in the model by parameter $\lambda$, is always higher than 1).

However, as opposed to those authors, we have considered a consumption index set up by differentiated goods and by a homogeneous good. Thus, it is possible to highlight the influence of the *demand* on the economic growth and give the consumers the importance that they actually hold in the economy of the nations.

We can summarise the model's predictions as follows - the economy's growth is all the faster as, *ceteris paribus:*

- The economy's investment in human capital is higher;
- The economy's investment in unskilled labour is lower (assuming that there is a low substitution elasticity of the inputs in the production of the homogeneous good);
- The R&D productivity is higher;
- The economic agents are more patient;
- The dimension of innovation is higher (the products' improvement *"ladder"*);
- The weight of the differentiated goods on the consumption index is higher.

## 5. References

Aghion, P.; Howitt, P. A model of growth through creative destruction. *Econometrica*, v. 60, n. 2, p. 323-351, 1992.

Barro, R. Economic growth in a cross section of countries. *Quarterly Journal of Economics*, v.106, n. 2, p. 407-443.

Barro, R.J. Are government bonds net wealth? *Journal of Political Economy*, v. 81, n. 6, p. 1095-1117, 1974.

Barro, R.J.; Sala-i-Martin, X. *Economic growth*, New York: McGraw-Hill, Advanced Series in Economics, 1995.

Becker, G.S.; Murphy, K.M.; Tamura, R. Human capital, fertility, and economic growth. *Journal of Political Economy*, v. 98, n. 5, p. S12-S37, 1990.

Bianchi, M. Novelty, Preferences and Fashion: When Goods are Unsettling. *Journal of Economic Behavior and Organization*, v. 47, p. 1-18, 2002.

Caballero, R.J.; Jaffe, A.B. How high are the giants shoulders: an empirical assessment of knowledge spillovers and creative destruction in a model of economic growth. *National Bureau of Economic Research Macroeconomics Annual Meeting*, p.15-74, 1993.

Collins, S.M. Lessons from Korean economic growth. *The American Economic Review*, v. 80, n. 2, p. 104-107, 1990.

Dixit, A. K. *Optimization in economic theory*. 2nd Edition, New York: Oxford University Press, 1990.

Dutt, A. K. Aggregate Demand, Aggregate Supply and Economic Growth. *International Review of Applied Economics*, v. 20, n. 3, p. 319–336, July 2006.

Fan, C.S. Quality, trade, and growth. *Journal of Economic Behavior & Organization*, v. 55, p. 271–291, 2004.

Fatas-Villafranca, F.; Saura-Bacaicoa, D. Understanding the demand-side of economic change: a contribution to formal evolutionary theorizing. *Econ. Innov. New Techn.*, v. 13, n. 8, p. 695–716, 2004.

GEPIE. *Inovação na indústria portuguesa: observatório MIE*, Abril 1992.

Grossman, G.M.; Helpman, E. Endogenous Innovation in the theory of growth. *Journal of Economic Perspectives*, v. 8, n. 1, p. 23-44, 1994.

Grossman, G.M.; Helpman, E. *Innovation and growth in the global economy*, Cambridge, Mass.; and London England: MIT Press, 2ª impressão, 1993.

Grossman, G.M.; Helpman, E. Quality ladders and product cycles. *Quarterly Journal of Economics*, v. 106, n. 2, p. 557-586, 1991a.

Grossman, G.M.; Helpman, E. Quality ladders in the theory of growth. *Review of Economic Studies*, v. 58, p. 43-61, 1991b.

Helpman, E. Endogenous macroeconomic growth theory. *European Economic Review*, v. 36, p. 237-267, 1992.

Lucas, R.E. On the mechanics of economic development. *Journal of Monetary Economics*, v. 22, p.3-42, 1988.

Lucas, R.E. Why doesn't capital flow from rich to poor countries? *The American Economic Review*, v. 80, n. 2, p. 92-96, 1990.

Lucas, R.E. Making a miracle. *Econometrica*, v. 61, n. 2, p. 251-272, 1993.

Mangasarian, O.L. Sufficient conditions for the optimal control of nonlinear systems. *SIAM Journal of Control*, v. 4, p. 139-152, 1966.

Metcalfe, S. Consumption, preferences, and the evolutionary agenda. *Journal of Evolutionary Economics*, v. 11, p. 37–58, 2001.

Pasinetti L. *Structural change and economic growth*. Cambridge University Press, Cambridge, 1981.

Pasinetti L. *Structural economic dynamics*. Cambridge University Press, Cambridge, 1993.

Prontryagin, L.S.; Boltyanskii; Gramkrelidze; Mischenko *The mathematical theory of optimal processes*. New York: Inter-science Publishers, 1962.

Ramsey, F. A mathematical theory of saving. *Economic Journal*, v. 38, p. 543-559, 1928; in *Readings in the modern theory of cconomic growth*, Editado por J.E. Stiglitz; H. Uzawa, The MIT Press, 5th. Ed., 1979.

Romer, P. Growth based on increasing returns due to specialization. *The American Economic Review, v. 77*, p. 56-62, 1987.

Romer, P.M. Endogenous technological change. *Journal of Political Economy*, v. 98, n. 5, pt.2, p. S71-S101, 1990.

Santos, F.B. *Cálculo das probabilidades*. Plátano Editora SA, 5ª Edição, 1988.

Saviotti P.; Mani GS. Competition, variety and technological evolution: A replicator dynamics model. *Journal of Evolutionary Economics*, v. 5, p. 369–392, 1995.

Saviotti P. Information, variety and entropy in technoeconomic development. *Research Policy*, v. 17, p. 89–103, 1988.

Saviotti P. *Technological evolution, variety and the economy*. Edward Elgar, Aldershot, 1996.

Saviotti P. The role of Variety in economic and technological development. In: Saviotti P., Metcalfe JS (eds) *Evolutionary theories of economic and technological change: present state and future prospects*, p. 172–208. Harwood Publishers, Reading, 1991.

Saviotti P. Variety, economic and technological development. In: Shionoya Y, Perlman M. (eds) *Technology, industries and institutions: studies in Schumpeterian perspectives*. The University of Michigan Press, Ann Arbor, 1994.

Saviotti, P. Variety, growth and demand. *Journal of Evolutionary Economics*, v. 11, p. 119–142, 2001.

Schultz, T. Investment in human capital. *The American Economic Review*, v. 51, n. 1, p. 1-17, 1961.

Schumpeter, J.A. *The theory of economic development*. Cambridge MA, Harvard University Press, 1934.

Segerstrom, P.S. Innovation, imitation, and economic growth. *Journal of Political Economy*, v. 99, n. 4, p. 807-827, 1991.

Silva, S.T. and Teixeira, A.A.C. On the divergence of evolutionary research paths in the past fifty years: a comprehensive bibliometric account. *Journal of Evolutionary Economics*, v. 19, n. 5, p. 605-642, 2009

Solow, R. M. A contribution to the theory of economic growth. *Quarterly Journal of Economics*, v. 70, p. 65–94, 1956.

Sonobe T., Hu, D., Otsuka, K. From inferior to superior products: an inquiry into the Wenzhou model of industrial development in China. *Journal of Comparative Economics*, v. 32, p.542-563, 2004.

Teixeira, A.A.C. *Capital Humano e Capacidade de Inovação. Contributos para o estudo do crescimento Económico Português, 1960-1991*, Série Estudos e Documentos, Conselho Económico e Social, Lisboa, 1999.

Witt, U. Economic Growth – What Happens on the Demand Side? Introduction. *Journal of Evolutionary Economics*, v. 11, p. 1–5, 2001a.

Witt, U. Learning to consume – A theory of wants and the growth of demand. *Journal of Evolutionary Economics*, v. 11, p. 23–36, 2001b.

# Part 2

# Measuring Technological Change

# 5

# Measuring Technological Change – Concept, Methods, and Implications

Byoung Soo Kim

*Korea Institute of S&T Evaluation and Planning*
*Republic of Korea (South Korea)*

## 1. Introduction

Technological change can be understood in terms of technological evolution. Because of this evolutionary aspect, a number of scholars have made analogies between technological innovation and biological evolution (Ziman, 2000). As technological advancement has played a crucial role in industrial development, almost every nation is concerned with monitoring technological change. Technology policies such as research and development (R&D), technology planning, technology management, etc. are related to this concern.

As accurately diagnosing the current situation is the first step in solving problems, measuring the current technological state is also an important stage in the making of technology policies. Indeed, there have been many measurement cases on the national level such as census surveys. As Porter argues, quantification is "a social technology" and "a crucial agency for managing people and nature" (Porter, 1995: 49-50). Quantitative measurement conducted by governments has as its aim the control of resources. Likewise, measuring the technological state on the national level aims to increase the efficiency of the national innovation system.[1]

There has been a long history of accurate measurement being valued by governments. Regarding systems of measurement driven by nation states, France tried to make an objective and accurate standard during the Enlightenment. The French government created the metric system as a result and this had complex effects on the political economy (Heilbron, 1990; Alder, 1995). The tradition of measurement was also important in Victorian Britain. Accurate measurement was a crucial value in such areas as imperial triumphs (Schaffer, 1995).

Besides national measurement cases, there has been a great deal of literature dealing with the analysis of technological states. For example, Merton (1978)[2] analysed the state of science and technology in seventeenth century England using available data. This kind of literature is an analysis of just a certain country. The measurement of technology in this chapter is

---

[1] The national innovation system is defined differently by different scholars. The system generally includes "all important economic, social, political, organizational, institutional and other factors that influence the development, diffusion and use of innovations" (Edquist, 2005: 183, as cited in Ediquist, 1997).

[2] This book was originally published in 1938 as Volume 4, Part 2 of Osiris.

focused on comparisons among nations rather than the absolute technological state of any given country.

In this chapter, I will deal with the theme of measuring technological change as measuring a technological state. Though there are various actors, areas and levels, I will focus on measuring activities relevant to the public sector. The importance of the private sector in terms of R&D is now greater than ever before, but governmental R&D policies still have great influence on technological development in the national innovation system. There is important context in the case of each nation. I will survey measuring methods and cases from a global perspective. However, I will also provide some specific empirical cases as needed.

More specifically, I will review the concept and context of measuring a technological state by using a historical and theoretical approach. Then, I will survey various methods of measurement with brief notes about different cases. The limitations of each measuring method will be illustrated as well. Finally, I will discuss the implications of measuring a technological state in terms of effectiveness and applicability.

## 2. Conceptual overview

Feller (2003) explains the evaluation of science and technology programs by using the newspaper reporter's algorithm of what, where, when, who, why and how. This format can easily communicate a difficult concept to readers. I have adopted his approach to introduce basic concepts of measuring technological change in this section.

### 2.1 What is it?

Measuring technological change is literally to measure the state of a certain technology along a changing path. As technology changes dynamically at all times, it is difficult to follow and check the moving state. Instead, a technological state can be measured at each stage. It is more realistic to describe measuring technological change as measuring technological states. As a result, these two expressions are sometimes used interchangeably in this chapter.

Measurement has different but overlapped meaning with similar concepts such as evaluation, assessment, and appraisal. As there are not strict distinctions among them, even experts sometimes mix them. However, evaluation is generally used for programs. R&D program evaluation exemplified this kind of activity.

Assessment is frequently mixed with evaluation. However, assessment has a somewhat different and specific meaning in 'technology assessment.' Technology assessment has been generally used to estimate the impacts, influences, or consequences of a specific technology on society and nature (Seki, 1992). Nowadays, it is also defined as "a scientific, interactive and communicative process which aims to contribute to the formation of public and political opinion on societal aspects of science and technology" by European institutes for technology assessment (Decker & Ladikas, 2004: 14).

The Office of Technology Assessment in the U.S. Congress, the first official organization for technology assessment in the world, had conducted the role of assessing technology from 1973 to 1995. The role of this office can be expressed as attempting to "minimize the negative effects of new technologies and maximize the positive effects" (Bimber, 1996: 26).

Unlike evaluation or assessment, measurement is to analyse an object in detail, precisely, and objectively more so than with other similar terms in the context of this chapter. Measuring a technological state is to estimate the precise level, degree, or stage of the technological trajectory. The results of measuring activities can be communicated as numbers, probabilities, or any other numerical figures.

## 2.2 When does it happen?

A technological state can be measured at each stage of the technological trajectory. If a certain technology is introduced, it evolves along its trajectory. As emerging technologies are intrinsically uncertain as to whether they will develop or fail in the future, each government has tried to measure the state of emerging technologies in their early stages. Governments make policies based on the results of this measurement.

Measuring a technological state is conducted periodically or irregularly by agencies. The results of these periodical measurements are used for time-series analysis in terms of technological change. On the contrary, irregular or one-time measurements are conducted for a specific technology. For example, the South Korean government conducts the measurement of the national technological level biennially, and conducts irregular measurement of specific technologies.

International organizations, such as the Organization for Economic Co-operation and Development (OECD), have regularly published the results of measuring technological states in terms of R&D and innovation. As a result of these regular reports relating to technological states, policy makers can understand the changing trends of technological states by time-series analysis.

## 2.3 Who does it?

There are various actors that initiate the measuring of technological states: nations, international organizations, companies, and so on.[3] In the public sector, governments and international organizations are the main actors. In particular, governments have been interested in measuring their resources. Much like current measurements, German-speaking states around 1800 already measured the "strength of the state" by calculating population and wealth (Nikolow, 2001: 23-24).

Though governments initiate measuring policies, the actual actors of the measurement are mainly agencies. This measurement is done by government agencies and private bodies as well. However, the government is ultimately the main actor as it is the contractor. Although companies do conduct technology measurement on their own, governments contract agencies for measuring work. Overall, there are more measurement results revealed by governments than companies. The latter are often unwilling to make public the results that they use for their private decision making. As a result of this context, I will focus on technology measurement in the public sphere, as I have mentioned.

---

[3] Likewise, Godin (2005) classifies participants in science measurement systems as following: (1) transnational organzations such as the OECD, UNESCO, EU, etc.; (2) national statistics agencies; (3) government departments; (4) organizations relating to S&T like the NSF; (5) university researchers; and (6) private firms.

As the importance of technology planning activities[4] has increased, the sphere of technology planning has become specialized in recent decades. The competency of an agency engaged in this work depends on the capabilities of its practitioners. These practitioners usually conduct other technology planning activities, as well as technology measurement. They possess the necessary methodology, practice, and experience in the area of technology measurement. However, they sometimes need other experts' knowledge in specific technology areas. Thus, technology measurement is conducted by many actors such as government officials, practitioners in agencies, expert groups including professors or engineers, and so on at the micro level.

## 2.4 Why do they do it?

Historically, developing countries such as South Korea and China have tried to catch up technologically to advanced nations. During such a 'catching-up' phase of a national innovation system, any developing country is willing to compare the state of their technologies to that of other nations. At this stage, the results of technology measurement can be a useful reference for catching-up strategies. The results of the measurement are generally used as references for R&D policies and innovation strategies by policy makers.

Technology measurement utilized exclusively by developing countries. Indeed, the United States, Japan, and advanced European countries have continuously conducted technology measurement. Even technologically advanced countries need to compare themselves to their competitors' states. The U.S. has also compared its domestic technological capability to that of other nations such as Japan, the EU, and so on. For instance, U.S. agencies published *Science Indicators 1972* to compare its level of science and technology in terms of R&D expenditure, R&D man power, technology transfer, etc. to that of other nations (National Science Board, 1973; Elkana et al., 1978). Sometimes the U.S. has officially compared its technological capability to a specific nation, such as Japan (Arrison et al., 1992).

The ultimate aim of technology measurement is to provide referential materials for decision making. Governments always need evidence for any decision making. As the trend of 'evidence-based policy' has deepened, the need for reference data has increased. Especially in the R&D area, in the late Twentieth Century, governments have invested their budget in various programs rather than evaluating the performance of their R&D. However, many governments try to evaluate the performance and results of their R&D programs these days. The 'science of science and innovation policy' initiated by the United States is an example of this policy change. Under the U.S. federal government's 'science of science policy' initiatives, agencies should develop tools to improve R&D portfolio management and better assess the impact and performance that results from their investments (Fealing et al., 2011).

## 2.5 How is it conducted?

Unlike other objects, technology is abstract and intangible. It is difficult for even experts to measure or estimate a technological state. Indeed, technology measurement is to estimate the technological state at a specific stage. There are some methods for estimating a

---

[4] In this chapter, the meaning of technology planning activities include technology foresight, technology measurement, technology assessment, technology roadmapping, and so on.

technological state piece by piece. Along the spectrum of available methods, there is a range between complexity and convenience. The more complicated a method is, the less practically attainable it is. Technology measurement is different from other methods generally used in laboratories or workshops. As mentioned above, the sphere of technology measurement has become specialized. There are some peculiar methods to measure the abstract state of technology. Specific types and methods will be explained in the next section.

## 3. Types and methods

In this section, I will classify measuring methods into five types: scoring models, data analyses, surveys, growth models, and indicators. Though each type of measuring is independent, different methods are sometimes combined.

### 3.1 Scoring models

Scoring models have been used as a means of ranking or rating technology quantitatively. In scoring models, detailed technological properties should be quantitatively measurable. There are various scoring models, but I will introduce the generally used models in this section. According to Martino's model, the technological state, in terms of total score, can be calculated by using the following equation. Each capital letter is a factor that composes the technological state. $A$ and $B$ are overriding factors. $(C, D, E)$, $(F, G)$, and $(I, J)$ are exchangeable factors within brackets. $I$, $J$, and $K$ are costs or undesirable factors (Martino, 1992).

$$\text{Score} = \frac{A^a B^b \left(cC + dD + eE\right)^z \left(fF + gG\right)^y \left(1 + hH\right)^x}{\left(iI + jJ\right)^w \left(1 + kK\right)^v} \tag{1}$$

$$(c + d + e = 1, \quad f + g = 1, \quad i + j = 1, \quad a + b + z + y + x = 1, \quad w + v = 1)$$

Gordon and Munson (1981) introduced a convention for measuring the state of the art. They suggested that different experts in the same technological area should estimate the state of the art at the same level. In the following convention, $P_n$ is the value of the $n^{th}$ parameter and $K_n$ is the weight.

$$\text{SOA} = \frac{P1}{P'1}\left[K_2 \frac{P2}{P'2} + K_3 \frac{P3}{P'3} + \cdots + K_n \frac{Pn}{P'n}\right] \tag{2}$$

The advantage of a scoring model is its quantitative measuring of results. As the results are calculated from detailed various factors or parameters, this method can well reflect technological properties. In spite of this advantage, the intrinsic aim of the scoring model cannot be easily realized. A scoring model should meet several requirements to fulfil its aim. The factors should be measurable and representative of the state of the art and data for measurement should be available. However, the concept of technology is too abstract to classify easily. It is also a difficult job to collect available data.

Recently, scoring models have not been used as a main method for measuring technological state. Instead, models have been complementarily used with other methods. For example, the Korea Institute of S&T Evaluation and Planning (KISTEP) has regularly measured the level of major technologies on behalf of the South Korean government since 1999. The KISTEP has also used Gordon's model to aggregate weighted values gained from data or surveys.

## 3.2 Data analysis

Data analysis is a measuring method using data from published papers and patents. This method is useful for measuring specific technologies that can be classified according to given standards. Patents are directly linked to technological performance. As papers are a category of the results from technological output, papers are also important resources for measuring a technological state in addition to patents.

There are a number of standard tools used to calculate the output of papers and patents. Citation analysis is a representative index used to measure a technological level qualitatively. As methodologies have developed, various indices such as the RCI (Relative Citation Index), CII (Current Impact Index), TII (Technology Impact Index), etc. are generally applied by practitioners. For example, recent cases such as Choi & Kim (2011) and Kim (2011) analyzed papers and patents in terms of publication numbers, citation numbers, citation index, specialization index, etc. Though they measured the state of special technologies in South Korea, they used data from SCI publications and U. S. patent publications to enhance the objectivity of the result.

Data analysis has emerged as an independent disciplinary area in other fields such as scientometrics, bibliometrics, informetrics, webometrics, netometrics, cybermetrics, and so on (De Bellis, 2009). Data analysis is more popular than other measuring methods. As the results of data analysis can be illustrated by quantified numbers, policy makers usually regard them as more objective than other figures. However, as many pieces of literature have shown, there are several issues such as language bias, timeliness of the analysis, comparability of the different research systems, statistical credibility, comparability of peer review judgement, and so on (Geisler, 2000; van Rann, 2004).

As papers and patents are just a portion of R&D outputs, there are some limitations in measuring the overall state of technology. Thus, data analysis should be used complementarily (Kim, 2010b). Output-centered measurement is not always appropriate for technology. In the case of a technology in the early or middle stages of its trajectory, outputs in these stages cannot reflect the overall potential of the technology. For example, five of the most important communications technologies such as telegraph, telephone, phonograph, personal computer, and cable television were not attractive to the market in the early stages of their development (Nye, 2006).

Nowadays, there are many cases of network analysis and mapping used in many agencies. Though using the same data, network analysis and mapping are different from data analysis. The most different aspect of the former is visualization. For instance, citation networks and distribution can be visualized as a map. Then, we can estimate technological change through comparing maps representing different times in terms of paper (or patent) citation. There are also limitations that apply to network analysis and mapping. Networking and mapping work are based on keywords, as in other forms of data analysis. Keyword setting is a time-demanding work and should be iterated until plausible results are achieved.

## 3.3 Surveys

The survey method is relatively easier than other methods. The basic assumption of the survey method is that experts are very well acquainted with the state of technology. In other words, the survey method utilizes the tacit knowledge of experts. Among survey methods, the Delphi survey has recently been used for measuring technology by many agencies. The Delphi survey was originally one of the foresight methods. In the early 1970s, the Japanese government initiated large-scale foresight surveys using the Delphi method, and the surveys have been repeated approximately every five years. In the Delphi survey, numerous experts are repeatedly surveyed with identical questions, with the results of previous rounds fed back to the respondents in order to revise their answers and draw out a consensus.

The U.S. had also used the Delphi method to solicit and synthesize the opinions and judgments of expert communities. For instance, the U.S. National Science Board conducted a Delphi survey in 1972. The topics of the survey were the future role of science and technology, the impacts of R&D funding, technological innovation, basic research, financial resource allocation, and graduate education (National Science Board, 1973). Though the Delphi survey was not directly conducted for measuring the technological state, the methodology was similar to the current measuring style used for this purpose. Moreover, the topics of the survey could be related to the technological state.

Since the late 1980s, the Delphi surveys have been popularized as a useful form of technology forecasting and measuring methods in many other countries. Particularly, many agencies in South Korea have applied the Delphi surveys to technology measurement since the turn of the century. Questionnaires in Delphi surveys usually include questions relating to the technological state such as the present level, impact, capability, competitiveness, and so on. The Delphi survey is a useful method in the case of a large number of technology areas, within a short time frame. Because of this strong point, the KISTEP has applied the Delphi survey for technology level evaluation since 1999. The agency has developed this method combined with a growth model in 2008 and 2010.

Survey methods are intrinsically subjective and qualitative. Though the answers of participants in the Delphi survey can be fed back to one another in repeated rounds, their decisions cannot be objective and quantifiable. Thus, there have been controversies whether the survey results are reliable. Some results from data analysis or indicators such as patent citation level can be objective and quantitative, but subjective methods can be employed in setting and drawing up policy. As Woudenberg (1991) reviewed, the Delphi method is not a science but an art of creating a consensus. The survey results are best understood as synthesized tacit knowledge and experts' intuition (Kim, 2010b).

## 3.4 Growth models

Growth models are not general methods for measuring something. Though they are more appropriate for forecasting rather than measuring, growth models are sometimes used for technology measuring. If other measuring methods are static, growth models are relatively dynamic. As growth models describe the changing trends of a technological state, policymakers can not only understand technological change dynamically, but also forecast the future state from growth curves.

In this section, S-shaped curve is used in the same sense as growth model. There are various S-shaped curves such as Bass, Pearl, Gompertz, and so on. The Bass model is an S-shaped curve model developed in the early stage. Nowadays, Pearl (3) and Gompertz (4) are generally used as S-shaped curve models. In both equations, $L$ means the upper limit of technological development.

$$Y(t) = \frac{L}{1 + \alpha e^{-\beta t}} \tag{3}$$

$$Y(t) = Le^{-\beta e - \alpha t} \tag{4}$$

While S-shaped curves provide us with a model of the dynamic change of technology, they require several assumptions (Porter et al., 1991; Martino, 1993; Kim, 2010a). These assumptions include a correct equation, a definite upper limit, proper fitting, and so on. Among them, the definite upper limit is not easily defined because the upper limit is a rather theoretical concept.

Recently, growth models have been applied for measuring technological level (Bark, 2007; Kim, 2010a, Kim, 2010b, Ryu & Byeon, 2011). For instance, the KISTEP applied the model to evaluate national core technologies in 2008 and 2010. Though there were limitations such as the problem of the upper limit, they are appraised as advanced cases in terms of measuring dynamic technologies.

Policymakers may prefer growth models because of their forecasting aspect. Technology forecasting is not also an easy job as there are numerous variables relating to technological change. The future is intrinsically uncertain, thus technology forecasting from an S-shaped curve is uncertain as well. Because the growth model is a form of trend analysis as well as an extrapolation method, there are intrinsic limitations of the unpredictability and fallibility. Technologies do not always follow forecasted trajectories due to unexpected factors such as 'disruptive innovation' (Christensen, 1997).

## 3.5 Indicators

There are many indicators and indices that can be used to measure a technological state. Though many indicators and indices are being made by various agencies, they are sometimes contrived as there are not enough objective criteria used for setting them. Sometimes, in the process of selection and rejection of indicators, political interests can be involved.[5] To give an example from among the globally accepted indicators, it is useful to review those used by the OECD. The OECD has also published reports relating to indicators in terms of innovation. Though technology and innovation are conceptually different, measuring indicators are sometimes mingled with each other, as the technological state usually results from innovation.[6] Among them, the Frascati and Oslo Manuals are guidelines for the measurement of scientific and technological indicators.

---

[5] Even in similar indicators, there are many results that differ according to which agency arranges them. Generally, there is a tendency to select indicators that are favorable to agency itself.
[6] As various scholars have conceptualized, the term innovation covers a vast set of changes including inventive and technological changes (Scerri, 2006; Colecchia, 2007).

The Frascati Manual is a set of guidelines for R&D statistics. The manual has been revised six times since it was issued in 1963. As R&D statistics are some of the most important indicators of economic growth, in terms of technological change, the Frascati guidelines have played a crucial role for many statistics and scoreboards made by the OECD and various countries. As a result, the Frascati guidelines have become a global standard for R&D surveys and statistics not only in OECD member countries, but also other organizations such as the UNESCO, the European Union and so on.

The Oslo Manual has been issued three times since 1992 and is more focused on innovation activities. The first edition, in 1992, included the results of surveys to develop and collect data on the process of innovation. The second edition, in 1997, updated its framework to enlarge the concept of innovation and the range of industries studied, and also improved innovation indicators to make them comparable among OECD countries. The most recent edition, issued in 2005, included a large amount of data and information from various surveys. It expanded its innovation measurement framework to relevant firms, services, etc. and types of innovation to organizational and marketing innovation.

Neither manual is a directly citable indicator for technological change. However, they provide relatively objective criteria for R&D and innovation statistics and indicators. Actually, many reliable technological indicators are based on data and information guided by those manuals. They are useful for comparing nations in terms of standardized indicators, but they have limitations regarding measuring specific areas of technology. As the guidelines deal with a relatively broad range of criteria for science and technology, more specific data and information would be needed in the case of measuring a specific technology such as nanotechnology.

There are also other indicators globally accepted such as IMD World Competitiveness. Though the IMD indicators have some sub-indicators such as scientific infrastructure, technological infrastructure, etc., the indicators are not sufficient to decide the technological state in each nation. Locally, there are some indicators such as Composite Science and Technology Innovation Index (COSTII, South Korea), Japanese Science and Technology Indicators (Japan), and so on. Though these annual indicators are published by each nation, they include information pertaining to other competitive nations such as the U.S., Germany, the U. K., China, etc. Thus, those indicators can be understood as criteria for comparative analysis among major nations.

## 4. Discussions

As I have briefly reviewed above, measuring a technological state is not easy but is instead a complicated activity. There are still controversies over measurement matters both theoretically and practically. Indeed, even the object of measurement is different in each case. As the concept of technology is abstract, a specific technology is sometimes regarded as an end product or the process of realizing the product as well. However, there is no universal consensus or standard that can be applied to measuring activities. Thus, a measurement designer should define meticulously the concept and scope of technologies as objects in the early stage of the activities.

In the stage of defining technology, the classification of detailed technology is also important, as well as the concept and scope of the object. Although there are some standardized classifications that apply to the global or national level, these standards cannot be applied to specific measurement cases. The standards are usually too broad to apply to more specific measurement. For example, the OECD standards based on the Frascati Manual do not include detailed technologies. For an example on the national level, although the National S&T Standards Classification System South Korea is somewhat more detailed than global ones, it is sometimes not compatible with the aim of specific practical measurement. General classification standards are based on normal disciplinary areas, but practical cases need specific classifications for their own aims. Each measurement project usually involves a separate process to classify the object technology, and thus there are often different criteria used to measure, even in similar technology areas.

Once the scope and classification of the object has been defined, practitioners should decide which method is proper to be applied. As explained above, there are various methods such as scoring models, data analysis, surveys, growth models, indicators, etc. for measuring a technological state. However, the present trend of measurement has tended to become heterogeneous and complex. For example, the KISTEP has continuously used a heterogeneous method in which the Delphi survey, growth model, scoring model, and data analysis are mixed for its technology level evaluation. As part of this effort, in 2008, the KISTEP developed a method that combines the Delphi survey with a growth model (Kim, 2010a; Kim, 2010b; Kim & Kim, 2010; Ryu & Byeon, 2011). Gordon and Munson's model was applied to calculate the results of the Delphi survey. Thus, data analysis was conducted complementally with the survey.

Though practitioners control the whole process of measurement, they need the expertise of specialists in specific technology areas. As mentioned above, experts' judgement is a kind of tacit knowledge. Their expertise can be converged in some ways including seminars, interviews, surveys, and so on. A greater number of experts can participate in surveys than in the other methods. However, the results of surveys are not always reliable. According to the results of the Delphi survey for the Marine S&T evaluation in South Korea, conducted by the KISTEP in 2010, the expertise of the participants in the survey was sometimes unstable, indeterminate and uncertain.

Measurement results are different according to the context of each country. For example, some countries prefer the ranking or grading of the results. In case of South Korea, the government prefers measurement results in terms of percentages. In this case, 100% represents the level of the most advanced country, as well as the criterion of measuring the level of South Korea. In other countries, like the United States or Japan, agencies usually publish their measurement results in terms of grading. Their grading is generally ranked according to five categories in terms of Likert scales.

As Godin (2005) argues, statistics reflect values, interests, and ideologies, so the results of measuring a technological state reflect such social factors that are involved in selecting indicators and publishing the results as well. Like any other scientific results, the results of measuring a technological state are also constructed by various human and non-human actors.[7] The measurement results are constructed by interacting relations among practitioners, technology specialists, measurement tools, figures, numbers, graphics, texts, and so on.

---

[7] This is the general perspective of the ANT (Actor-Network Theory) introduced by Bruno Latour, Michel Callon, and John Law. The constructing processes of making scientific results is well shown in Latour's books (Latour & Woolgar, 1979 ; Latour, 1987).

Once the results are published, the results in terms of numbers and figures have their own trajectory like a living thing. Published figures and numbers are sometimes appropriated by users. Policy makers use them to assess R&D programs, to compare R&D capabilities among countries, to make technology roadmaps, and so on. They are frequently cited as evidence in various official documents. Even if some mistakes are found in the figures and numbers, it is usually difficult to revise them after they have been made official. Though they are well aware of the limitations of each measurement results, policy makers usually use the results in order to legitimize their policies in terms of evidence.

Technology measurement sometimes plays an important role in visualizing previously vague technologies as concrete ones. Intangible technology can be visualized with the help of texts, numbers, graphs or maps. In other words, such figures and numbers represent intangible technology. The results represent experts' judgement on technology in terms of tacit and explicit knowledge. Though limitations mentioned above exist, the results of measurement have generally been used to help shape relevant policies. The necessity and applicability of measuring technological states has been generally accepted as well. Thus, more sophisticated methodologies are needed in the field of technology measurement.

## 5. References

Alder, K. (1995). A Revolution to Measure: The Political Economy of the Metric System in France, In: *The Values of Precision*, M. Norton Wise (Ed.), 39-71, Princeton University Press, ISBN 0-691-03759-0, Princeton, New Jersey, the United States

Arrison, T. S. et al. (Eds.) (1992). *Japan's Growing Technological Capability: Implications for the U.S. Economy*, National Academy Press, ISBN 0-309-58481-7, Washington D.C., the United States

Bark, P. (2007). A Theoretical Approach and Its Application for a Dynamic Method of Estimating and Analyzing Science and Technology Levels: Case Application to Ten Core Technologies for the Next Generation Growth Engine, In: *Journal of Korea Innovation Society*, Vol. 10, No. 4, ISSN 1598-2912, pp. 654-686 (Written in Korean)

Bimber, B. (1996). *The Politics of Expertise in Congress: The Rise and Fall of the Office of Technology Assessment*, State University of New York Press, ISBN 0-7914-3059-6, Albany, New York, the United States

Choi, M & Kim, B. S. (2011), Technology Level Evaluation in Agricultural Science and Technology, In: *The PICMET 2011 Conference Bulletin*, 84, Portland State University, Portland, Oregon, the United States

Colecchia, A. (2007). Looking Ahead: What Implications for STI Indicator Development?, In: *Science, Technology and Innovation Indicators in a Changing World: Responding to Policy Needs*, OECD (Ed.), 285-298, ISBN 978-92-64-03965-0, OECD, France

Christensn, C. M. (1997). *The Innovator's Dilemma: When New Technologies Cause Great Firms to Fail*, ISBN 0-87584-585-1, Harvard Business School Press, Cambridge, Massachusetts, the United States

De Bellis, N. (2009). *Bibliometrics and Citation Analysis: From the Science Citation Index to Cybermetrics*, Quantitative Information Study Forum, (Trans.), KISTI, ISBN 978-89-6211-476-8 93020, Seoul, Korea (Korean translation)

Decker, M. & Ladikas, M. (Eds.) (2004). *Bridges between Science, Society and Policy: Technology Assessment-Methods and Impacts*, Springer, ISBN 3-540-21283-3, Berlin, Germany

Edquist, C. (2005). Systems of Innovation: Perspectives and Challenges, In: *The Oxford Handbook of Innovation*, Jan Fagerberg et al. (Eds.), 181-208, ISBN 0-19-926455-4, Oxford University Press, New York, the United States

Elkana, Y. et al. (1978). *Toward A Metric of Science: The Advent of Science Indicators*, ISBN 0-471-98435-3, John Willey & Sons, the United States

Feller, I. (2003). The Academic Policy Analyst as Reporter: The Who, What and How of Evaluating Science and Technology Programs, In: *Learning from Science and Technology Policy Evaluation: Experiences from the United States and Europe*, P. Shapira & S. Kuhlmann, (Eds.) 18-31, Edward Elgar, ISBN 1-84064-875-9, Cheltenham, the United Kingdom

Fealing, K. H. et al. (Eds.) (2011). *The Science of Science Policy: A Handbook*, Stanford Business Books, ISBN 978-0-8047-7078-1, Stanford, California, the United States

Geisler, E. (2000). *The Metrics of Science and Technology*, Quorum Books, ISBN 1-56720-213-6, Westport, Connecticut, the United States

Gläser, J. & Laudel, G. (2007). The Social Construction of Bibliometric Evaluations, In: *The Changing Governance of the Sciences: The Advent of Research Evaluation Systems*, R. Whitley & J. Gläser, (Eds.) 101-123, Springer, ISBN 978-1-4020-6745-7, Dordrecht, The Netherlands

Godin, B. (2005). *Measurement and Statistics on Science and Technology: 1920 to the present*, Routledge, ISBN 0-415-32849-7, New York, the United States

Gordon, T. J. & Munson, T. R. (1981). A Proposed Convention for Measuring the State of the Art of Products or Processes, In: *Technological Forecasting & Social Change*, 20, ISSN 0040-1625, pp. 1-26

Heibron, J.L. (1990). The Measure of Enlightenment, In: *The Quantifying Spirit in the 18th Century*, Tore Frängsmyr et al. (Eds.), 207-241, University of California Press, ISBN 0-520-07022-4, California, the United States

IMD (2011). *IMD World Competitiveness Yearbook 2011*, IMD World Competitiveness Center, ISBN-13 978-2-9700514-5-9, Lausanne, Switzerland.

Kim, B. S. (2010a). *A Study on the Dynamic Method of Estimating Technology Levels Based on the Technology Growth Model*, Korea Institute of S&T Evaluation and Planning, Research Paper Series 2010-23, Seoul, Korea (Written in Korean)

Kim, B. S. (2010b). A Case of Forecast-Based Technology Evaluation and Its Implications, In: *International Journal of Technology Intelligence and Planning*, Vol. 6, No. 4, ISSN 1740-2840, pp. 317–325

Kim, B. S. & Kim, J. H. (2010). Integrating Technology Planning Methods for R&D Strategy in Maritime Technology, In: *The 3rd ISPIM Innovation Symposium Book of Abstracts*, K.R.E. Huizingh et al. (Eds.), 3, ISBN 978-952-214-005-4

Kim, B. S. (2011). *A Study on Establishing a Master Plan for the National Marine Environment R&D Program*, Ministry of Land Transport and Maritime, Gwacheon, Korea (Written in Korean)

Latour, B. (1987). *Science in Action: How to Follow Scientists and Engineers Through Society*, Harvard University Press, ISBN 0-674-79291-2, Cambridge, Massachusetts, the United States

Latour, B. & Steve Woolgar (1979). *Laboratory Life: The Construction of Scientific Facts*, Sage Publications, ISBN 0-691-09418-7, Beverly Hills, California, the United States

Martino, J. P. (1993). *Technological Forecasting for Decision Making*, McGraw-Hill, ISBN 0-07-040777-0, the United States

Merton, R. K. (1978). *Science, Technology and Society in Seventeenth Century England*, Humanities Press, ISBN 0-85527-357-7, New Jersey, the United States

National Science Board (1973). *Science Indicators 1972*, National Science Foundation, Washington D.C., the United States

Nikolow, S. (2001). A. F. W. Crome's Measurements of the "Strength of the State": Statistical Representations in Central Europe around 1800, In: *The Age of Economic Measurement*, J. L. Klein & M. S. Morgan, (Eds.) 23-56, Duke University Press, ISBN 0-8223-6517-0, the United States

Nye, D. E. (2006). *Technology Matters: Questions to Live With*, The MIT Press, ISBN 0-262-14093-4, Cambridge, Massachusetts, the United States

OECD (2002). *Frascati Manual: Proposed Standard Practice for Surveys on Research and Experimental Development*, OECD, ISBN 92-64-19903-9, Paris, France

OECD (2005). *Oslo Manual: Guidelines for Collecting and Interpreting Innovation Data*, Third Edition, OECD, ISBN 92-64-01308-3, Paris, France

Porter, A. L. et al. (1991). *Forecasting and Management of Technology*, John Willey & Sons, ISBN 0-471-51223-0, the United States

Porter, T. M. (1995). *Trust in Numbers: The Pursuit of Objectivity in Science and Public Life*, Princeton University Press, ISBN 0-691-03776-0, Princeton, New Jersey, the United States

Ryu, J. & Byeon, S. C. (2011). Technology level evaluation methodology based on the technology growth curve, In: *Technological Forecasting & Social Change*, 78, ISSN 0040-1625, pp. 1049-1059

Schaffer, S. (1995). Accurate Measurement is an English Science, In: *The Values of Precision*, M. Norton Wise (Ed.), 135-172, Princeton University Press, ISBN 0-691-03759-0, Princeton, New Jersey, the United States

Scerri, M. (2006). The Conceptual Fluidity of National Innovation Systems: Implications for innovation measures, In: *Measuring Innovation in OECD and non-OECD countries*, William Blankley, Mario Scerri, Neo Moloija & Imraan Saloojee (Eds.), 9-19, HSRC Press, ISBN 0-7969-2062-1, Cape Town, South Africa

Seki, S. (1992). What Can We Learn from Technology Assessment?, In: *Japan's Growing Technological Capability: Implications for the U.S. Economy*, Arrison, Thomas S. et al. & Committee on Japan, National Research Council (Eds.), 43-56, National Academy Press, ISBN 0-309-58481-7, Washington D.C., the United States

van Raan, A. F. J. (2004). Measuring Science: Capita Selecta of Current Main Issues, In: *Handbook of Quantitative Science and Technology Research: The Use of Publication and Patent Statistics in Studies of S&T Systems*, H. F. Moed & W. Glänzel, (Eds.) 19-50, Kluwer Academic Publishers, ISBN 1-4020-2702-8, Dordrecht, The Netherlands

Woudenberg, F. (1991). An Evaluation of Delphi, In: *Technological Forecasting & Social Change*, 40, ISSN 0040-1625, pp. 131-150

Ziman, J. (Ed.) (2000). *Technological Innovation as an Evolutionary Process*, Cambridge University Press, ISBN 0-521-62361-8, Cambridge, the United Kingdom

# 6

# Quantitative Technology Forecasting Techniques

Steven R. Walk
*Old Dominion University*
*USA*

## 1. Introduction

Projecting technology performance and evolution has been improving over the years. Reliable quantitative forecasting methods have been developed that project the growth, diffusion, and performance of technology in time, including projecting technology substitutions, saturation levels, and performance improvements. These forecasts can be applied at the early stages of technology planning to better predict future technology performance, assure the successful selection of new technology, and to improve technology management overall.

Often what is regarded as a technology forecast is, in essence, simply conjecture, or guessing (albeit intelligent guessing perhaps based on statistical inferences) and usually made by extrapolating recent trends into the future, with perhaps some subjective insight added. Typically, the accuracy of such predictions falls rapidly with distance in time. Quantitative technology forecasting (QTF), on the other hand, includes the study of historic data to identify one of or a combination of several demonstrated technology diffusion or substitution patterns. In the same manner that quantitative models of physical phenomena provide excellent predictions of systems behavior, so do QTF models provide reliable technological performance trajectories.

In practice, a quantitative technology forecast is completed to ascertain with confidence when the projected performance of a technology or system of technologies will occur. Such projections provide reliable time-referenced information when considering cost and performance trade-offs in maintaining, replacing, or migrating a technology, component, or system.

Quantitative technology forecasting includes the study of historic data to identify one of or a combination of several recognized universal technology diffusion or substitution patterns. This chapter introduces various quantitative technology forecasting techniques, discusses how forecasts are conducted, and illustrates their practical use through sample applications.

## 2. Introduction to quantitative technology forecasting

Quantitative technology forecasting is the process of projecting in time the intersection of human activity and technological capabilities using quantitative methods. For the purposes of forecasting, technology is defined as any human creation that provides a compelling advantage to sustain or improve that creation, such as materials, methods, or systems that displace, support, amplify, or enable human activity in meeting human needs. It will be shown how rates of new technology adoption and rates of change in technology performance take on certain characteristic patterns in time.

A quantitative technology forecast includes the study of historic data to identify one of several common technology diffusion or substitution trends. Patterns to be identified include constant percentage rates of change (such as the so-called "Moore's Law"), logistic growth, logistic substitution, performance envelopes, anthropological invariants, lead/lag (precursor) relationships, and other phenomena. These quantitative projections have proven accurate in modeling and simulating technological and social change in thousands of applications as diverse as consumer electronics and carbon-based primary fuels, on time scales covering only months to spanning centuries.

Invariant, or at least well-bounded, human individual and social behavior, and selected (genetic) human drives underlie technology stasis as well as change. In essence, humans and technology co-evolve in an ecosystem that includes the local environment, our internal physiology, and technology (where the technology can be considered external or complementary physiology). The fundamental reliability of quantitative technology forecasts is being supported by ongoing developments in modeling and simulation derived from systems theory, including complex adaptive systems and other systems of systems research.

Carrying out a quantitative technology forecast includes selecting a technology of interest, gathering historic data related to changes in or adoption of that technology, identifying candidate "compelling advantages" that appear to be drivers of the technology change, and comparing the rate of technology change over time against recognized characteristic patterns of technology change and diffusion. Once a classic pattern is identified, a reliable projection of technology change can be made and appropriate action taken to plan for or meet specific technology function or performance objectives.

QTF as defined here, as it seeks to determine the 'fit' of time-stamped growth or diffusion of technological data to ubiquitous yet mathematically simple models, does not include probabilistic, non-temporal based, or other relational methods that are seeing increased use in data-mining and data visualization efforts in determining technological and social trends. Many commercial products are now available that perform statistical and other algorithmic analyses among data in large databases to determine otherwise indiscernible relationships. Such analyses can be useful, for example, in marketing and sales, business intelligence, and other activities requiring a better understanding of relationships among systems of complex interactions among components or agents, and the system or individual response to change. While these practices do include observing or trending change over time, the analyses usually involve only secondarily linear temporal projections including statistically based measures of uncertainty or risk. Moreover, the focus of these methods is most often understanding or visualizing static or cause-effect relationships, rather than understanding primarily the growth, diffusion, substitution, etc., which are primary foci of the highly temporal-based QTF methods.

## 2.1 Methodologies

Quantitative technology forecasting has been applied successfully across a broad range of technologies including communications, energy, medicine, transportation, and many other areas.

A quantitative technology forecast will include the study of historic data to identify one of or a combination of several recognized universal technology diffusion or substitution trends. Rates of new technology adoption and rates of change of technology performance characteristics often can be modeled using one of only a relatively small number of common patterns. The discovery of such a pattern indicates that a fundamental diffusion trajectory, envelope curve, or other common pattern has been found and that reliable forecasts then can be made.

The quantitative forecasting techniques are "explanatory principles" (Bateson 1977), that is, sufficient by their reliability for the purposes of modeling technology diffusion patterns and forecasting technology adoption. Many researchers have attempted to develop fundamental theories underlying substantiate the commonly found patterns, such as extending theories of system kinematics and other advanced systems theories, to varying success and acceptance in the field. The ubiquity of the various patterns has been studied also using systems theory and complexity modeling, such as the complex adaptive systems approach.

### 2.1.1 Logistic growth projection

Forecasters had their first significant successes in predicting technological change when they used exponential models to project new technological and social change (e.g., Malthus, 1798, as cited in ). It was deemed only logical that a new technology at first would be selected by one, than perhaps two others, and these people in turn, two others each, and so on, in a pattern of exponential growth. Ultimately however, as in any natural system, a limit or bound on total selections would be reached, leading early researchers to the use of the logistic (or so-called S-curve) to model technological and social change.

In the late 20th Century, researchers in the United States such as Lenz (Lenz, 1985), Martino (Martino, 1972, 1973), and Vanston (Vanston, 1988), and others around the world, such as the very prolific Marchetti (Marchetti 1977, 1994, 1996) refined forecasting methods and showed that the logistic model was an excellent construct for forecasting technological change. The logistic displayed virtually universal application for modelling technology adoption, as well as for modeling effectively many other individual and social behaviors.

The classical logistic curve is given by:

$$P(t) = \kappa / \{1 + \exp[-\alpha(t - \beta)]\} \tag{1}$$

This simple three-point curve is defined by $\kappa$, the asymptotic maximum, often called the carrying capacity; $\alpha$, the rate of change of growth; and $\beta$, the inflection point or mid-point of the curve. Figure 1 illustrates the idealized logistic curve of technology adoption or diffusion.

A popular means to visualize the growth match to the ideal logistic curve is by way of a linear transformation of the data. The Fisher-Pry transform (Fisher and Pry 1971) is given by:

$$P'(t) = F(t) / [1 - F(t)], \text{ where } F(t) = P(t)/\kappa \tag{2}$$

where $F(t)$ is the fraction of growth at time t, given by

$$F(t) = P(t)/\kappa \tag{3}$$

The Fisher-Pry transform projects the ratio of per unit complete and per unit remaining of a growth variable.

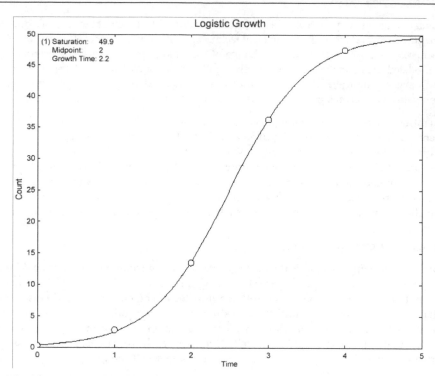

Fig. 1. Ideal logistic growth curve (Adapted from Meyer et al, 1998).

Figure 1 illustrates the idealized logistic curve of technology adoption or diffusion. Figure 2 shows the logistic growth of the supertanker of maritime fleets presented in a popular format developed by Fisher and Pry (Fisher and Pry 1971) that renders the logistic curve linear. Figure 3 shows the growth pattern of a computer virus that infected computers on worldwide networks.

Note that the time and level of saturation (peaking) of the logistic trajectory is a key indicator of change: it can signal the emergence of new or substitute technology.

## 2.1.2 Constant rate of change (performance envelope)

Technology change occurs within dynamic and complex systems of human behavior. The growth and diffusion of technology influences and is influenced by the activities of humans as individuals and groups at varying scales. The adoption of new technology requires intellectual, material, energy, and other resources to be redirected, increased, and otherwise managed as required in the implementation of the new technology.

When a new technology emerges having the substantive compelling advantage such that it will successfully substitute for an incumbent technology at some higher, but still (physiologically complementary) practical performance level, humans in groups tend to go about the changeover in a methodical way, managing to maintain equilibrium in the vast array of a culture's interaction and interdependent social, material, and economic systems.

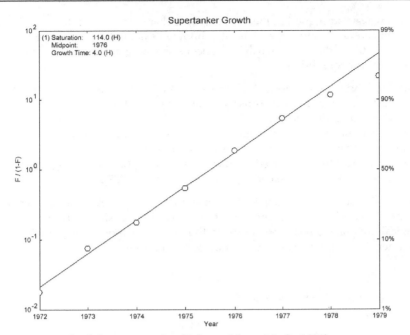

Fig. 2. Logistic growth of the supertanker (Adapted from Modis 1992).

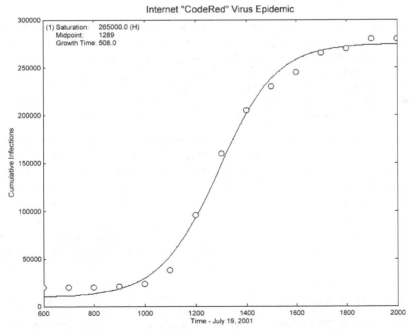

Fig. 3. Logistic growth of a network computer virus (Data from Danyliw and Householder, 2001).

The result suggests strongly that the adoption and change of substitute technologies is far from random and rarely sudden, and usually follows a smooth transition, at a rate of change dependent on the either consciously or unconsciously maintained by individual and collective forces for equilibrium.

This result contradicts theories of "disruptive technology". It is true that individuals or subgroups can face significant disruption when a new technology arrives and displaces, especially if rapidly, the incumbent technology. However, QTF research suggests that careful study can unveil the technological or physiological parameter that governs or paces the technological changeover, reflected in a smooth trajectory of the change characteristic on a larger scale. Marchetti commented, "Show me the disrupting technology, and I will find you the logistic curve", (personal communication, February 2007).

Forecasters call the curve of sequential performance levels of adopted technologies a *performance characteristic curve*, and search for its telltale shape in the history of a technological area of interest. The nature of the curve is exponential growth, where, if a quantity $x$ depends exponentially on time $t$ the growth expression is

$$x(t) = a^{b/\tau} \tag{4}$$

where the constant $a$ is the initial value of $x$,

$$x(0) = a \tag{5}$$

and the constant $b$ is a positive growth factor, and $\tau$ is the time required for $x$ to increase by a factor of $b$.

Figure 4 shows an example of the performance characteristic curve for transistor density on a microprocessor chip, the popular "Moore's Law". Intel CEO Gordon Moore could target the timing of the introduction of new technology performance and thereby target R&D efforts to a horizon performance. The performance envelope shows that the successful emergence and integration of a new technology can be predicted inversely, i.e., *the performance envelope provides a future order of merit that an incubating technology must achieve*. In other words, the horizon drives the technology, not the other way around.

Figure 5 shows the performance envelope history of primary industrial energy conversion. Note that the performance trajectory technique identifies the horizon of expected technological performance, in this case, an energy conversion efficiency of about 75% in 2050. Along this trajectory, we see that only fuel cell technologies, of all emerging energy conversion technologies, are capable of meeting the 2050 efficiency horizon.

Core performance envelopes have been identified in many industries, and likely exist for all industries and technology areas. The performance envelope is a valuable index and decision tool for technology planning, including technology watch and horizon scanning activities.

### 2.1.3 Logistic substitution

Transitions from one technology or performance level to the next tend to follow neat, manageable patterns. In the 1960's, Fisher and Pry (Fisher and Pry 1971) analyzed hundreds of technological substitutions in history and devised a method to graph the substitution

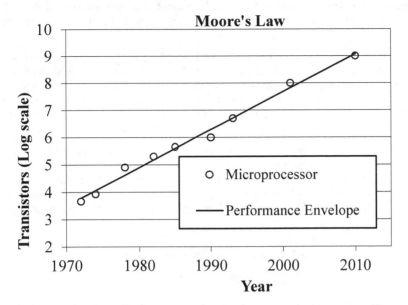

Fig. 4. Moore's Law - Performance envelope of microchip transistor density (Data from Intel Corp. 2001)

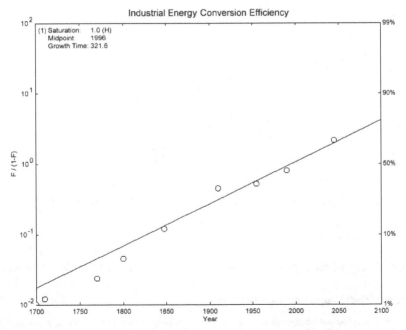

Fig. 5. Performance envelope of industrial energy conversion technology, projection to 2050 (Adapted from Ausubel and Marchetti, 1997).

patterns in linear form. The curve-fit is more easily estimated by observation of the straight line of the Fisher-Pry transform as compared to estimating fit along the sweeping logistic curve.

Figure 6 illustrates a typical logistic substitution pattern, here the substitution of automobiles for horses as the preferred 'personal vehicle' for transportation. Studies have identified the logistic substitution pattern in technologies as diverse as substitutions in fiber optic transmission networks (Vanston 2008) and the substitution of latex for oil-based paints (Modis 1992). In the maritime industry there evolved a multiple-pulse logistic substitution of motor-over-steam-over-sail in ship propulsion technology (Figure 7).

Being able to predict the emergence and diffusion of substitute (often disruptive, on local scales) technology is a powerful tool in any comprehensive, strategic technology watch program. Knowing the trajectory of an overcoming technology enables reliable projections of critical time horizons of technology change. QTF Logistic Substitution provides a straightforward method to illuminate these evolving patterns of current and future events.

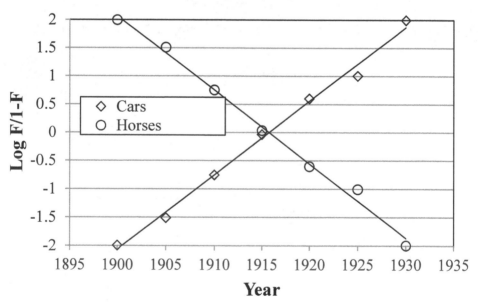

Fig. 6. Logistic substitution of primary personal travel mode (Fisher-Pry display format).

### 2.1.4 Precursor (lead-lag) growth relationship

The implementation or adoption of a technology has been shown to vary logistically. When one technology is dependent on or otherwise closely related to a previous development, the two trajectories are found to synchronize in a steady lead-lag relationship (see Figure 8).

## US Maritime Propulsion Technology Substitutions

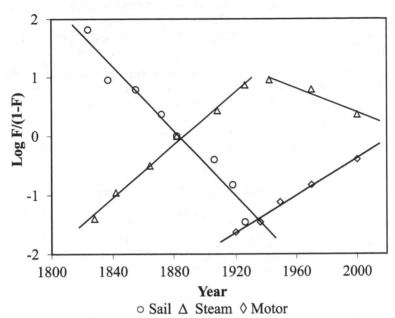

○ Sail △ Steam ◇ Motor

Fig. 7. Substitution of US maritime propulsion technology (Adapted from Modis 1992).

## Typical Technology Precusror Relationship

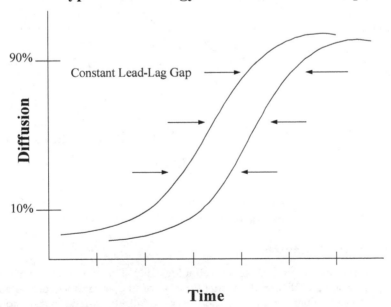

Fig. 8. Constant lead-lag logistic relationship.

Studies have shown that the worldwide discovery of petroleum resources has led the production of oil by a fixed period over many decades (Modis 2000). Studies have shown also that the diffusion in the USA industry of networked desktop personal computers followed the same shape logistic trajectory as the precursor technology, stand-alone PCs (Poitras and Hodges 1996).

Figure 9 shows the lead-lag relationship between patents and research publications in quantum dot technology. Academic publications and US patent office databases were mined to capture the two parallel logistic growth pulses. Notice again the nearly constant 4-year time lag from beginning to end of the pulses. It would have been readily forecast around the year 2000 that within six more years the early patent pulse would end and quantum dot technology would next evolve into broad applications and commercial viability.

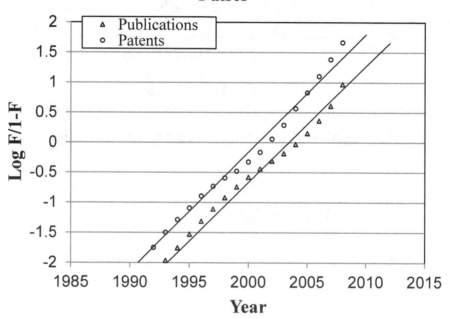

Fig. 9. Approximate constant 2-year precursor relationship between quantum dot technology patents and publications (Data from Walk 2011b).

### 2.1.5 Anthropological invariants

In the grand history of the progression of technological change, one of the striking results is evidence, otherwise not identified or identifiable, of the invariance of certain human behaviors. While technologies offer many and perhaps infinite varieties of how to get things done, the things humans do want to get done, generally, have remained the same for hundreds and thousands, and perhaps millions of years. For example, travel and communication patterns, depicted in broad averages of commuting or foraging times, or in numbers of human exchanges, have been shown to be constant across time and cultures.

In the case of the commuter, the average commute in the United States has remained at about one half hour since the automobile became the main choice of personal mobility a century ago. The advice to automobile manufacturers is that seat design need only accommodate the average drive, about 30 minutes. No matter how much the manufacturer's investment, no matter how much more advertised, the average user is going to drive the automobile a half hour per day, out and back.

The compelling advantages in the design and performance of technologies can be viewed as artifacts of unchanging human behavioral preferences. As an example, Figure 10 shows the more or less constant acceptable (and, by complex social feedback mechanisms, so engineered and designed) risk of death by automobile in the United States, over nearly an entire century.

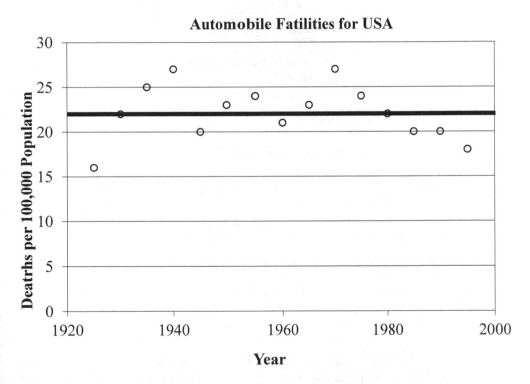

Fig. 10. Risk of having a fatal automobile accident in the US (Adapted from Marchetti 1994).

## 2.2 Sample published QTF applications

Thousands of studies using QTF techniques to project or monitor technological change have resulted in identifying the underlying logistic, performance envelope, substitution, or other ubiquitous fundamental trajectory. Table 1 provides a very short list of sample published studies.

Sections 2.2.1 and 2.2.2 provide further examples of QTF studies with more in depth discussion of the process of understanding and extracting strategic meaning from study results.

| QTF Technique | Application | Publication Source |
|---|---|---|
| Logistic Growth | Aluminum in Automobiles | Bright 1973 |
| | Commercial Space Launches | Walk 2011a |
| Logistic Substitution | Popular Recording Media | Meyer, et al 1999 |
| | World Primary Energy Substitution | Marchetti 2000 |
| Performance Envelope | Hard Drive Density | Christenson 1997 |
| | Internet Bandwidth | Nielsen 2011 |
| Precursors | Oil Discovery and Use | Modis 2000 |
| | FORTRAN | Walk 2011b |
| Anthropological Invariant | Age of world shipping fleet casualties | Walk 2004 |
| | Share price of DJIA stock | Modis 1992 |

Table 1. Sample QTF published studies

## 2.2.1 Human travel: Wanderlust, exploration, and settlements

Marchetti published a remarkable series of quantified technological studies of the locomotive habits of humans (Marchetti 1994). Modis (Modis 1992) provides a very interesting graph of the 'discovery' of the Americas, of which Figure 11 is an adaptation. While many interesting insights flow from this remarkably simple set of sailings across the Atlantic, the reader's attention is called to the fundamentally logistic growth in probing the New World.

The proceeding graphs (Figures 12 to 14) show human exploration patterns of our sailings across space: to the Moon, Venus, and Mars. The Mars probes were plotted using a 5-year running average to smooth the clusters of probes launched when Mars and Earth were in optimal launch positions. Note the two distinct pulses of Mars probe activity, the second approaching 90% of saturation at the time of writing this chapter. These various space probe logistic patterns would be nearly interchangeable if we were to normalize scales.

At the end of the logistic saturation in Columbus' era, in the early 1700's, 'permanent' settlements in the 'new' world had been established and commercial trade was in early flourish. The 'new' world was much like the 'old' world in almost every aspect of habitability, our legacy of inherited survival needs could be met readily, and so we relocated.

The logistic pattern of discovery, of which the probes to the 'new world' are an excellent example, is repeated in space explorations. However, the probes have not been followed by the settlements and commercial trade phase.

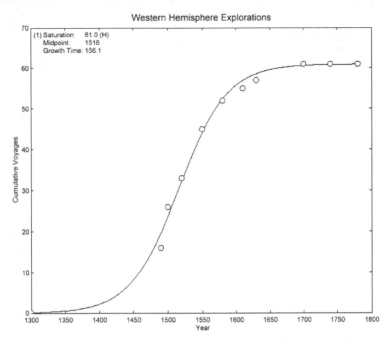

Fig. 11. Logistic pattern of discovery voyages of the Western Hemisphere (Adapted from Modis 1992).

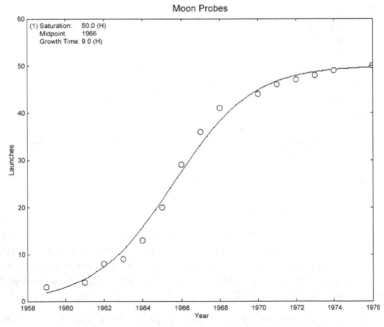

Fig. 12. Logistic growth of Moon Probes (Data from NASA 1958-1976)

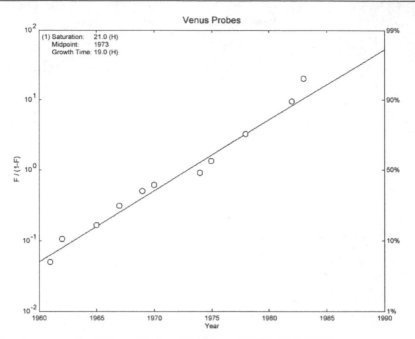

Fig. 13. Logistic growth of Venus Probes (Data from NASA 1960-1990)

Let us consider, in this example, what the consistent patterns in world and solar system probes might tell us. Among many things, the studies tell us what is obvious, but that can be lost in heady visions of technology promise: we simply do not inhabit places we do not like after having visited them. QTF studies have shown there exist identified and unidentified, i.e., conscious or unconscious, invariants in human behavior apparently lain down by millions of years of gene-coded and cultural evolution that we cannot override by our wishing, will, or law. For example, after exploring the far polar regions, by air, on land, and under the sea, people have not settled there. The landscapes simply are too inhospitable, the criteria of hospitableness being a short list of bounded behaviors and needs, some identified in the history of technological adoption and change. We have not built cities under the sea, as was projected by futurists and popular media in the late 20[th]century, and QTF perspectives indicate it is extremely unlikely we ever will.

We might, then, consider seriously that we have reached the end of our Moon, Venus, and Mars explorations, and that the idea of anyone ever settling either Moon, Mars, or Venus, is a wish fading from our collective mind.

The results of this sample QTF study suggest a larger perspective on space travel. Humans have learned to travel the ocean, and even penetrate a short distance beneath its surface to travel and meet other physiological (e.g., nutritional) needs, but we have not chosen to live below the ocean surface. We can look at this frontier geometrically, as that of a well-traveled outer spherical surface of a downwardly vast and uninhabitable ocean of water.

We can look at the space frontier by an analogous geometrical stretch of the imagination. Consider our very active and frequently travelled low earth orbit (LEO, up to 2000 km)

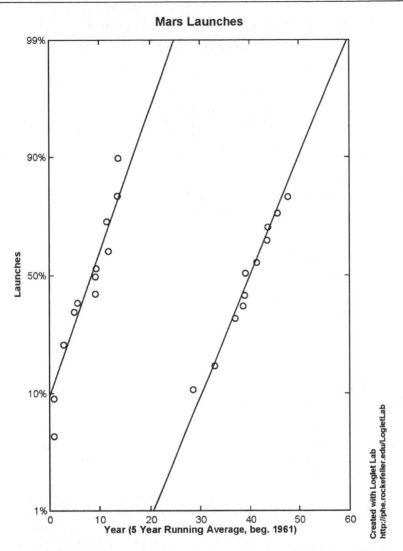

Fig. 14. Two-pulse logistic growth of Mars Probes (Data from NASA 1961-2011)

sphere as an interior spherical surface of an outwardly vast ocean of space. We have learned to travel this space ocean 'surface', and even penetrated to a short distance (in a radially outward direction, such as across our solar system). However, we appear to be balking on habitations to any 'depth' of space beyond LEO.

## 2.2.2 Sample of linked research and production trajectories

Figure 15 reveals an interesting and telling symmetry in the superposition of the logistic diffusions of supertankers and engineering publications related to supertankers. The initial publication citations began about 20 years before the commencement of commercial

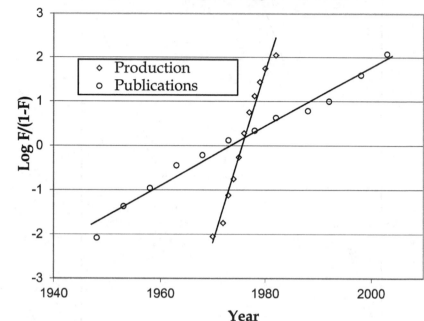

## Logistic Growth of Supertanker Production and Supertanker Engineering Publications

Fig. 15. Growth of research and production of supertankers (Production data from Modis 1992. Publication data by search of Compendex database.)

production, and ended about the same period following the cessation of production. The year of peak rates of growth of publications and production, the mid-points of the logistic curves, was the same. We wrote about them, built them, and wrote about them some more, in a tightly choreographed *pas de deux* of logistic movement.

Note the advantage of such information in technology watch and horizon scanning programs: knowing the relationship between publications and production can yield important timing intelligence for both trending technology and performance trajectories or diffusion in a technology watch program, and for timing of peak saturation and likely emergence of new technology in a horizon scanning program.

## 3. Typical QTF study procedures and practical considerations

The following tasks outline the typical procedure followed in a QTF study effort.

1.  Identify candidate forecast technologies for which the following criteria are met:
    *   The technology or technology performance requirement lies on a critical path in a strategic development plan or in a specific program requirement.
    *   Reliably accurate, time-referenced data of prior technology adoption rates or performance change rates are available quickly and at reasonable cost.
    *   The forecast can be prepared in no longer than three months

- The forecast will provide the opportunity to undertake follow-on forecasts, either by further increasing the forecast scope or scale in the selected technology, or by branching into related technologies or applications.
2. Select the best candidate technology or technology performance requirement for the project based on the selection criteria.
3. Gather experts in the history, application, or research and development of the selected technology or performance requirement.
4. Hold discussions that stress conceptual thinking to develop hypotheses of 1) the (single) performance criteria that led to the adoption, substitution, and overall evolution of the technology, and 2) the (fundamental) human utility driving the technology changes.
5. Obtain time-referenced data to test the hypotheses developed in the expert discussions.
6. Analyze the results of the forecasts.
   - Revisit, clarify, and reaffirm the rationale of the hypotheses underlying the forecast.
   - Corroborate the results, if possible, through the analysis of analogous or parallel technology change.
   - Consider future extreme external events that could significantly influence the forecast.
7. Repeat steps 1-6 for other candidate critical technologies, and learning from the outcomes of the forecasts, identify further opportunities for routine technology forecasting in future technology planning and development.

## 3.1 Determining growth metrics

Quantitative technology forecasts focus on changes in time of fundamental characteristics of technologies or systems of technologies. While the models and mathematics are relatively simple, time-based, quantitative constructs, expertise in the technology, social regime, or other discipline of the studied data is required to identify the index (metric or measure) that is the locus of change, and which becomes the variable on the vertical scale of the plots, and to assure the reliability of subsequent forecasts. A look back at the many sample plots in this chapter show a wide variety of indices. Often, the index or metric is not an intuitive measure. The reader is encouraged to review the indices identified in achieving the published results shown in Table 1 to observe the wide variation in metrics, and to note those metrics that at first blush might not be considered in initial trajectory analyses.

It is important to remember in all cases that *successful* technological innovation, technology change, and technology performance are linked fundamentally to relatively narrowly invariant behaviors or strong preferences in human needs and behavior. Candidate 'improved' or 'next-generation' technologies must match individual human or socially advantageous human needs and capabilities to survive. Too fast, too small, too big, and so forth, and the new technology will fail, i.e., not be adopted, even though it might provide significant 'improvement' in some performance measure. Technology can be said to be simply extended physiology, and those technologies that have too high a behavior penalty or too low a behavior advantage will not be chosen.

As an example of this phenomena, the commercial airline industry growth was found to follow a performance trajectory based on a metric including passenger seat and speed (Bright and Schoeman, 1973). The number of seats and cruising speeds were logical and intuitive candidates for metrics of airline performance change over time. However, taken individually, each variable yielded no common pattern over time. However, the product of the two yielded an impressive "Moore's Law" for commercial airlines, consistent over more than six decades of development. Considering the product further, it did appear that it did indeed represent a logical growth index. Little advantage is gained if only one of these variables is increased in airliner design, with a result of a lot of people travelling only slowly, or only a few people travelling but very fast. The combination makes intuitive sense, however, with more people going faster simultaneously and increasingly and at a steady rate of change, resulting in an age of low cost frequent travel for the majority of people living in the industrialized world.

Another substantiating and plausible characterization lies in the physics of the change in airliner performance: assuming increasing passenger seats means increasing size and thus mass (though not necessarily in a linear relationship), the trajectory in time can be seen as a continuous improvement in momentum: mass times velocity.

## 3.2 Curve-fitting and forecast assessment

The goal in a QTF study is to develop the best-fit logistic (or other) curve from the available data. Least-squares fit and other common curve-fitting and valuation methods apply. Researchers at Rockefeller University (Meyer et al, 1999) developed and have made available curve-fitting software that provides real-time interfaces to modify, for instance, logistic equation constants (midpoint, saturation level, and slope) to visualize the best fit curve. Almost any commercial modeling and simulation software program that allows custom equation entry and automatic curve-fitting will provide satisfactory means to develop and best-fit curves to data. Debecker and Modis (Debecker and Modis, 1994) have published a broad investigation of logistic curve development and have suggested quality indices to evaluate the reliability of projecting forward from a plot of limited available data. Meyer (Meyer 1996) also describes several means to derive confidence indices to assess the quality, and thus forecast reliability, of a curve-fit to data in the case of logistic modeling.

## 3.3 Advantages of QTF

Technology forecasting is receiving increasing interest in private industry and government agencies as leaders and decision makers consider the potential damage of so-called disruptive technologies and loss of competitive advantage in areas such as national defense and security. Many methods have been and are being developed to forecast technology change exclusive of QTF techniques. Some of the forecasting methods rely on drawing qualitative consensus from gathered expert opinion (Delphi method) or performing extensive probabilities-based numerical analyses on vast databases of accumulated information (complex domain forecasting) for example from social networks.

As a conceptual framework, QTF *avoids all* of the following problems associated with the expert and complex domain forecasting methods:

- Methods that are subjective, labor-intensive, reliant on qualitative analysis, un-scalable, un-integrated, and generally untested
- Approaches to technology forecasting are often narrow in scope, focus, and applicability

- Systems that ingest and store up to petabytes of information and require internal cleaning and manipulation before analysis
- Products that leverage social network analysis that do not perform entity disambiguation
- Technologies that analyze social data for trends usually focused on one type or source of data or focus on one subject area
- Technologies that predict events or trends built on proprietary approaches that cannot be independently scrutinized

In addition, QTF techniques *can meet* the following long-term needs as a practical conceptual framework:

- Can be automated, gathering archived and streaming data and self-generating and self-evaluation diffusion and performance patterns
- Can be applied as a unifying conceptual framework to address all types of questions regarding technology forecasting, including both technology watch and horizon scanning.

QTF requires only that the data or information have a time signature. Data can be mined from stored databases such as publications, government agency, and non-government (commercial, private, institutional, etc.) databases. Real-time or near-real time data is harnessable from Internet, social network, and other communications data streams to complement or supplement stored historic data or to analyze and track change on a finer, near-term, time scale.

## 3.4 Risks and challenges of QTF

Careful review is necessary of forecasts and trajectories or the identification of a potential anthropological invariants, and standard procedure calls for expert review of the data and metrics used in a study, as noted in Sections 3, 3.1, and 3.2. One risk and one challenge stand out as important and frequent hindrances to an organization's application, adoption, and acceptance of QTF techniques to replace or supplement expert opinion methods or deep data analysis approaches.

1. Resistance to new and less complex forecasting techniques despite persistent unreliability and maintainability of expert and probabilistic methods. Massive data mining and deep analysis programs are finding very wide diffusion at the time of this writing following certain success in their ability to identify data trends that can be leveraged for business or strategic advantage. The models are not necessarily designed for technology forecasting applications, but have some direct and indirect value. In comparison, QTF can appear to be a retrograde analytical approach, what with only 3-paremeter equations and the simplest mathematical growth curves. However potent or reliable, the QTF methods can be misjudged and overlooked.
2. Unavailability of time-stamped data in a critical area of interest. Data is not always stored in time-based data sets, or in performance indices or quantities of interest. Commercial data services are available but can be costly. Government-reported data usually is very useful, but is limited often to general information on public utilities such as transportation, communications, etc., data.

### 3.5 Commercially available and dedicated software

All the curves used in QTF studies can be created in any number of software packages, such as Microsoft's EXCEL, Parametric Technology's MATHCAD, or Mathwork's MATLAB. Several analytical packages have been developed specifically for technology forecasting and are available without charge, including Loglet1 and Loglet2 by the Program for the human Environment at Rockefeller University (http://phe.rockefeller.edu/LogletLab) and IIASA Logistic Substitution Model II by the International Institute for Applied Systems Analysis (http://www.iiasa.ac.at/Research/TNT/WEB/Software/LSM2/lsm2-index.html).

## 4. Ongoing research

Research continues to better understand the applicability of the logistic (or performance envelope) curves in diverse technological and social change phenomena. Synergies have evolved between QTF techniques and advances in modeling and simulation, especially in the areas of complex adaptive systems and evolutionary systems. The attempts are to derive more complex parametric models of technology and social change to validate the consistency of the heuristic and aggregate QTF models (see, for example, Linstone 2011, Eriksson 2008, and Könnölä, et al, 2007).

## 5. Conclusion

Quantitative technology forecasting includes the study of historic data to identify one of or a combination of several recognized universal technology diffusion or substitution patterns. This chapter has introduced various quantitative technology forecasting techniques, discussed how forecasts are conducted, and illustrated their practical use through sample applications.

QTF is founded on relatively simple mathematical models portraying diffusion or growth. Systems and systems of systems can exhibit very complex behavior yet result in the emergence of common trajectories of output or outcomes. QTF captures these emergent phenomena to characterize technological change in reliable, repeatable techniques embodied in logistic and performance envelope curves in time.

QTF techniques depend only on compilations of time-stamped data or information to develop reliable trajectories along the simple two- or three-parameter models. Broadly successful forecasts of diffusion and performance patterns have utilized such data as bibliometric, citation, authorship, and patent analyses, and activity such as in social networks, search engine patterns, government agency statistics databases, and many other sources of time-configured data and information. Additionally, while it might seem logical to utilize all available data to generate a QTF (e.g., all patents ever issued, or all papers ever published, on a particular technology topic), the power of the QTF methods is that they typically require only a sampling (or subset) of data and information to generate a valid curve as well as maintain, or update, that curve over time. This translates to significant savings on data and resources to implement strategic technology watch or horizon scanning activity.

A QTF-based technology forecasting framework can integrate readily the *quantitative* outcomes of *qualitative* forecasts produced, for example, by such methods as expert surveys,

Bayesian, agent-based, and other disparate types of analyses, thereby increasing the validity and reliability of the expert- or probabilistic-based projections. For example, while a single survey of subject matter experts might not add significant quantitative information to a certain technology projection, a *progression of surveys* over time can unveil the emergence, diffusion, and saturation level of technological ideas, attitudes, and valuations, etc. When incorporated in the QTF-based technology forecasting framework, such information helps create the full, reliable picture of the dynamics of the adoption – or rejection – of technological change.

# 6. References

Ausubel, J.H. and C. Marchetti, C. (1997), "*Elektron: Electrical Systems in Retrospect and Prospect*", *Technological Trajectories and the Human Environment*, J.H. Ausubel and H.D. Langford, eds., National Academy, Washington DC, pp. 115-140.

Bateson, G. (1977): *Steps Toward an Ecology of Mind*, Ballantine Books.

Bright, J., & Schoeman, M. (Eds.). (1973). A Guide to Practical Technological Forecasting, Prentice hall, 0-13-370536-6, Engelwoof Cliffs, NJ USA.

Bright, J., (1973), *A Guide to Technological Forecasting*, Prentice-Hall.

Christensen, C., (1997) The Innovator's Dilemma, Harvard Business Press.

Danyliw, R., & Householder, A. (2001): Adapted from data in http://www.cert.org/advisories/CA-2001-23.html. Last visited October 2011.

Debecker, A, & Modis, T. (1994) Determination of the uncertainties in S-curve logistic fits, Technological Forecasting and Social Change, Volume 46, Issue 2, June 1994, Pages 153-173

Eriksson, E.A., and Weber, K.M., (2008) "Adaptive Foresight: Navigating the Complex Landscape of Policy Strategies", *Technological Forecasting and Social Change* 75:4, pp. 462-482.

Fisher, J. C., & Pry, R. (1971), "A Simple Substitution Model for Technological Change", *Technology Forecasting and Social Change*, Vol.3, pp. 75-78.

Intel Corporation (2001): Data in the public domain in various trade publications.

Könnölä,T., Brummer,V., and Salo, A., (2007) "Diversity in foresight: Insights from the fostering of innovation ideas," Technological Forecasting and Social Change, Volume 74, Issue 5, 608-626.

Lenz, R. C. (1985), "A Heuristic Approach to Technology Measurement", *Technological Forecasting and Social Change*, Vol. 27, pp 249-264

Liberty Fund, Inc., 2000, "Malthus, An Essay on the Principle of Population: Library of Economics", Available from *EconLib.org*.

Linstone, H. (2011), "Three eras of technology foresight", Technovation Volume 31, Issues 2–3, 69–76

Marchetti, Cesare (1977): See, for example, "Primary Energy Substitution Models: On the Interaction Between Energy and Society", *Technological Forecasting and Social Change*, Vol. 10, pp. 75-88.

Marchetti, Cesare (1994): Adapted from data in "Anthropological Invariants in Travel Behavior", *Technology Forecasting and Social Change*, Vol. 47.

Marchetti, Cesare (1996), "Looking Forward – Looking Backward: A Very Simple Mathematical Model for very Complex Social Systems", paper presented at "Previsione Sociale e Previsione Politica", Urbino, Italy.

Marchetti, Cesare (2000), "On Decarbonization: Historically and Perspectively", HYDROFORUM 2000 Munich, 11-15 September 2000. http://www.cesaremarchetti.org/archive/electronic/ir-decarb.pdf. Last visited January 2012.

Martino, J. P., (1972), *Technological Forecasting for Decision Making*, Elsevier.

Martino, J. P. (1993), *Technological Forecasting for Decision Making*, 3rd Ed. McGraw-Hill, pp. 281-282.

Meyer, P., (1996) "Bi-Logistic Growth", *Technological Forecasting and Social Change*, Technological Forecasting and Social Change 47:89-102 (1994).

Meyer, P., Yung, J., & Ausubel, J., (1998): *Loglet Lab for Windows Tutorial*, Program for the Human Environment, Rockefeller University. The Loglet program was used for Figures 1,2, 3, 5, 6, 7, 11, 12, 13, 14, and 15.

Meyer, P., Yung, J., & Ausubel, J., (1999). "*A Primer on Logistic Growth and Substitution: The Mathematics of the Loglet La b Software*", Technological Forecasting and Social Change Vol 61(3), pp. 247-271.

Modis, T. (1992). *Predictions*, Simon and Schuster, New York..

Modis, T. (2000), "Natural Gas Replaces Oil", Growth-Dynamics Newsletter, Theodore Modis, publ. See http://www.growth-dynamics.com/news/Jul17.html. Last visited January 2012.

Nakicenovic, N. (1986). Adapted from data in *Technological Forecasting and Social Change*, Vol. 29.

NASA Launch Chronologies, http://nssdc.gsfc.nasa.gov/planetary/chronology.html. Last visited January 2012.

Nielsen, J. (2011), *Neilsen's Law of Internet Bandwidth*. http://www.useit.com/alertbox/980405.html. Last visited January 2012.

Poitras, A.J. & Hodges, R. L., (1996) *Computer Technology Trends, Analysis, and Forecasting*, Technology Futures, Inc., publ., Austin, Texas.

Vanston, J. H. (1988), *Technology Forecasting: An Aid to Effective Management*, Technology Futures, Inc., Austin, TX.

Vanston, L. K., (2008) "Forecasts for the US Telecommunications Network", http://www.telenor.com/no/resources/images/018-028_ForecastsUSTelecomNetworks-ver1_tcm26-36175.pdf. Last visited January 2012.

Walk, S. R. (2004), "*Quantitative Technology Forecasts of Select Maritime Technologies and Implications for Maritime Education and Training*", paper presented at 5th General Assembly of the International Association of Maritime Universities, Launceston, Tasmania.

Walk, S. R., (2011a), "Projecting Technology Change to Improve Space Technology Planning and Systems Management", *Acta Astronautica, Journal of the International Academy of Astronautics, 68-7, pp.853-861.*

Walk, S. R., (2011b), "Improving Technological Literacy Criteria Development through Quantitative Technology Forecasting", ASEE Annual Conference and Exposition, Vancouver, Canada.

# Part 3

# The Empirics of Technological Change

# 7

# Internationalization Approaches of the Automotive Innovation System – A Historical Perspective

António C. Moreira[1] and Ana Carolina Carvalho[2]
*[1]DEGEI, GOVCOPP, University of Aveiro*
*[2]ISCA, University of Aveiro*
*Portugal*

## 1. Introduction

Innovation has long been recognized as one of the main pre-requisites for the sustainable development of firms, regions and nations. It is the outcome of fruitful collaboration and interaction between business firms and a wide variety of institutional actors. This interaction is so important that a systemic approach was put forward in order to identify and deal with all relevant actors (Carlsson et al., 2002; Cooke & Memedovic, 2003).

The literature on innovation systems is embedded with different conceptualizations: National Innovation Systems (Lundvall, 1992; Nelson, 1993; Porter, 1990), Regional Innovation Systems (Cooke et al., 1997), Sectoral Innovation Systems (Malerba, 2002; Breschi & Malerba, 1997) and Technological Systems (Carlsson 1995; Carlsson et al., 1995); though used extensively, they are hard to operationalize (Rondé & Hussler, 2005).

The importance of Regional Innovation Systems (RIS) stems from the increasing interaction of regional actors on the outcome of the innovation process. Although they have similar economic and industrial backgrounds, RIS are far from homogeneous (Heidenreich, 2004). In a similar way, Innovation Systems need to take into account the different spatial and technological levels of their actors (Arocena & Sutz, 2000; Carlsson, et al., 2002).

Heidenreich (2004) argues that local experience-based, context-bound knowledge, trust-based patterns of cooperation and the path dependent accumulation of competencies are crucial for a RIS to prosper. Heidenreich (2004) has also found that the governance structure of a RIS may, to some degree, limit the innovation process of the region. Finally, substantial differences are expected between large and small regions in terms of economic performance and in the functioning of the RIS.

Taking into account that RIS stem from cooperation and the accumulation of path-dependent competencies, it is expected that small regions in less-favored countries – with a limited number of developed industries and with a historical governance structure of strong centralization patterns – to be naturally disadvantaged in terms of regional innovation systems.

Although sectoral-oriented studies have been quite disseminated, due to Porter's (1980) five-forces model dealing with the firm's external threats (competitors, substitutes, newcomers, purchasers and suppliers), they have normally been focused on understanding industry dynamics.

Departing from a systemic approach to innovation, Malerba (2002) managed to overcome classical sectoral studies and included the following actors:

- individuals (consumers, entrepreneurs, scientists, etc);
- firms (users, producers and input suppliers);
- organizations such as universities, research institutes, financial agents, syndications and technical associations;
- groups of organizations.

It is the interaction between these actors that generated a Sectoral Innovation System (SIS) structured around knowledge, technology, inputs and demand related to a specific economic sector. Although mediated by the national or regional system, due to its narrow focus the SIS is much more suited to study the sector's innovation, learning and production processes (Malerba, 2002).

Malerba (2004) used three dimensions to characterize and analyze sectoral systems of innovation: technological knowledge, actors or networks, and institutions.

Although the sectoral approach has been used to analyze innovation at a regional base (Moreira et al., 2007; 2008), the evolution of a sectoral system of innovation has not been characterized. As a consequence, one of the main objectives of this chapter is to characterize the historical evolution of the automotive innovation system in Portugal following a sectoral perspective.

Internationalization is a complex phenomenon that has been extensively researched in the last decades (Ruzzier et al., 2006). It has been defined in various ways (Benito & Welch, 1997; Chetty, 1999; Chetty & Hunt, 2003; Welch & Luostarinen, 1988; Calof e Beamish, 1995) according to several understandings.

In essence, internationalization could be understood as the process by which a firm starts developing its main operations abroad. Although the majority of studies have been focused on the firm perspective, this study will be focused on the automotive sector as a unit of analysis.

Due to the diversity of approaches, (Moreira, 2009) this study is mainly focused on the internalization theory (Morgan e Katsikeas, 1997), the transaction cost analysis (Williamson, 1975; Gilroy, 1993), the eclectic paradigm (Dunning, 1988) and the Uppsala model (Johanson & Wiederheim-Paul, 1975; Johanson & Vahlne, 1990).

Due to the economic importance of the automotive industry in Portugal, it is our intention to analyze the evolutionary perspective of internationalization. The objective is therefore, to analyze how the main theories of internationalization are related to the milestones of the automotive innovation system as well as, the factors that led to the internationalization of the auto industry.

In this sense, this analysis is focused on the six historical steps of the automotive industry, comprised between 1937 and 2011. In this manner, we seek to understand how the internationalization perspective of the sectoral system has been constrained over the years.

Methodologically, this chapter is based on historical data of the auto industry and is framed according to the sectoral innovation perspective and internationalization theories. Therefore, a descriptive approach is used as we try to match the main internationalization theories with the historical perspective of the auto industry.

The chapter is divided in six sections. After the introduction, the second section addresses the theoretical underpinning of innovation systems. The third section describes the main internationalization approaches. The fourth section describes the main historical evolution of the automotive industry. The fifth section, based on the concepts put forward in the last three sections, relates the automotive innovation system with the internationalization theories during the six most important phases of Portuguese automotive production. Finally, in the sixth section the main conclusions and future perspectives are drawn up.

## 2. Innovation systems

The term *National Innovation System* (NIS) was first mentioned by Freeman (1987) regarding the complexity and dynamics of the innovation process. Freeman (1987) defines NIS as a set of public and private institutions whose activities and interactions generate, import, change and diffuse new technologies.

An innovation system is composed of elements and relationships that interact in the production, diffusion and use of new knowledge (Lundvall, 1992). According to Nelson and Rosenberg (1993), the crucial part of an innovation system is the set of institutional actors that underpin a differentiating innovative performance.

Lundvall (1992) differentiated between *narrow* and *broad* innovation systems. The former approach identifies institutions that promote acquisition and dissemination of knowledge. The broad approach recognizes that innovations can be generated in every part of an economy and that practical, cultural and economic differences may determine the sources of innovation.

Although there is some controversy with the systemic perspective, the concept of NIS has been extensively used as an analytical tool by several important institutions such as, the European Commission, the OECD and the United Nations Conference on Trade and Development (UNCTAD). As a result of the growing number of articles, Edquist (1997) has put forward the concept of *Innovation Systems* based on the following features:

- Innovation, intrinsically linked to learning, which is at the center of the analysis;
- Its holistic and interdisciplinary perspective, as it comprises economic, institutional, organizational, social and political determinants;
- Its historical and path dependent perspective;
- The lack of an optimal system;
- The important role given to institutions and their linkages in the search of a systemic order;
- The interdependence among actors, which plays a key role in new knowledge creation process.

Edquist (1997) defends the relevance given to NIS, based on the fact that it captures important aspects of the policy of the innovation process. In fact, a NIS addresses governmental policies of science, technology and innovation, R&D competencies of both public and private institutions, educational systems and financial support institutions. The key issue of an innovation system is that its ability to generate innovations does not depend on how individual actors perform, but rather on how they interact (Gregersen & Johnson, 1996).

Although the initial analysis of innovation systems has been applied to a national reality, Cooke *et al.* (1997) carried out the same framework on a regional perspective giving rise to *Regional Innovation Systems* (RIS). As stated by Cooke & Morgan (1998), national innovation systems have been influenced by two different drivers: globalization and regionalization. Accordingly, and based on the fact that some regions have managed to prosper more than others, the thriving regions can become important development centers – as the actors as well as their linkages are important for generating and disseminating new tacit-based knowledge and specific knowledge spillovers, difficult to access outside the system – for capturing new foreign direct investment and thus foster the pervasive nature of innovation.

RIS and NIS are not all alike as they depend on the location and flow of knowledge between actors. Asheim & Isaksen (2002) have put forward three types of RIS: (a) the territorially embedded regional innovation networks; (b) the regional networked innovation system; and (c) the regionalized national innovation systems.

Geographical, social and cultural proximity play an important role in the generation of firms' innovative activities in territorially embedded regional innovation networks. Learning-by-doing and learning-by-interacting are the main knowledge generating mechanisms as the presence and interaction with knowledge providers is relatively modest (Asheim & Isaksen, 2002). Although knowledge flows interactively among actors, the probability of producing radical innovations is low due to the modest presence of knowledge providers.

In regional networked innovation systems, the regional institutional infrastructure is more systemic than in territorially embedded regional innovation networks (Asheim & Isaksen, 2002). As regional networked innovation systems are regarded as ideal-typical, RIS local firms have a higher probability of generating radical innovations than in the previous situation, which is a consequence of the strong networking activities of the regional cluster of firms.

In regionalized national innovation systems, outside actors are involved in regional firms' innovative activities. As a consequence, knowledge providers are located outside the region, which to some degree limits the innovation process of the region.

Asheim & Isaksen (2002) point out that in regionalized national innovation systems, the interaction between knowledge organizations and firms are based more on specific research work between knowledge providers outside the region and the local industry, than on integration and continuous involvement of all actors.

On a different note, Heidenreich (2004) introduced two types of RIS: the Entrepreneurial and the Institutional. The latter is characterized by an industrial structure with a strong position of low and medium technology, a governance structure dominated by formal, and in general, public institutions and, by a business structure characterized by the important role of multinational companies. The former, on the other hand, is characterized by a solid bed of strong knowledge-based SMEs, creative entrepreneurs in new technological fields and, by a strong position in knowledge intensive services.

Following a spatial, social, material and chronological dimension of regional innovation processes, Heindenreich (2004) addressed the different innovation dilemmas and regional governance of 15 regions. He put forward the following four dilemmas:

- Firms must take advantage of local experience-based, context-bound knowledge to face world-wide oriented competition;
- Trust-based patterns of cooperation between actors (firms, schools, R&D institutions, authorities and users) must facilitate the recombination of technical knowledge and the embeddedness of new technologies;
- The coupling of scientific, economic, political, technical and cultural actors in order to facilitate the reciprocal adjustment of perspectives and actions;
- The accumulation and path-dependent development of competencies in order to (re)generate regional competitiveness.

For Heidenreich (2004), these dilemmas pose different challenges to innovation governance. In the case of *Grassroots* regions, the main challenge is to overcome the highly fragile institutional order threatened by firms' individualistic behavior and weak local authorities. In *Dirigiste* regions, the main challenge is to overcome the stability of institutional order and to generate regional cooperation among actors. Finally, in *Networked* regions, the main challenge for governance structures is to maintain entrepreneurial interests and to match regional R&D, technology transfer and economic policy to the new global challenges of the knowledge-based economy.

Clearly, the specificity, complexity and interdependence of different RIS depend on both the technological knowledge of actors and the type of innovation system governance.

# 3. Theories of internationalization

The internationalization of firms can be analyzed according different theoretical approaches. Although there is no single approach to comprehend and explain the internationalization process, some of them might be used complementarily. As many theories have been put forward to explain firms' internationalization, this chapter only addresses four of the most important theories in international business.

## 3.1 Internalization approach

The internalization theory is centered on the notion that firms aspire to develop their own internal markets whenever transactions can be performed within the firm itself with the minimum possible cost and, that such transactions will be maintained as long as the internalization benefits are larger than the internalization costs (Buckley & Casson, 1993).

According to this theory, organizations internalize transactions that are inoperative or costly in the market by identifying an efficient mode of transfer that minimizes costs. Firms evolve from simple internationalization processes (exportation) to more complex ones in order to guarantee that shared resources and knowledge are kept inside the firm (Buckley & Casson, 1976).

This theory congregates all the studies that analyze the internationalization of firms from the point of view of the transaction cost economy (Williamson, 1975). A company is extended abroad when the transactions are internalized beyond national borders (Buckley & Casson, 1976; Casson, 1979) because the costs within the company are smaller than those in the market. Accordingly, an important part of this theory addresses "how" firms decide to internationalize. In this way, it is possible to compare the establishment of foreign branch offices (vertical integration) with other forms of internationalization such as licensing (Hennart, 1982, 1989).

Intangible assets, such as technology and know-how (Buckley & Casson, 1976; Teece, 1986) or specific assets (Anderson & Gatignon, 1986) play an essential role in the decision to internationalize as well as in the selection mode ("reason" and "how"). Technology and know-how endow firms with the necessary competences for entering new markets. Furthermore, firms protect themselves against opportunism from possible partners, internalizing international operations when asset specific resources are important for international competitiveness.

This positive relation between firm internationalization and the existence of intangible assets has been extensively supported (Buckley & Casson, 1976; Caves, 1982; Denekamp, 1995; Dunning, 1980; Pugel, 1978; Wolf, 1977). Some studies support a similar relation between intangible assets and the vertical integration of distribution channels in foreign markets (Anderson & Coughlan, 1987; John & Weitz, 1988; Klein et al., 1990.), as well as the integration of innovation activities (Caves, 1982; Kamien & Schwartz, 1982).

Internalization involves one form of vertical integration that brings new operations and activities, previously carried out in intermediate markets, under the control of the firm, specifically in imperfect markets (Ruzzier et al., 2006).

The internalization process is very attractive for firms that tend to desire a tighter control over their operations. This is especially important when firms try to exploit their competitive advantage based on technology and knowledge. Multinational companies can explore their advantages through franchising and licensing, for example, but the internalization of their activities allow them to maintain direct control of their assets and the dilution of their property.

### 3.2 The eclectic paradigm

Dunning's (1988) eclectic paradigm, also known as the OLI paradigm, is based on the internalization theory and seeks to explain the different forms of international production as well as the selection of countries for investing abroad.

According to Dunning (1988), the internationalization of economic activity is determined by the achievement of three kinds of advantages. Firstly, ownership advantages (O) which are firm specific and are related to the accumulation of intangible assets, technological capabilities or product innovations. Accordingly, firms operating in foreign markets take advantage of these essential competences to outperform their competitors in international markets.

Secondly, the location advantages (L) are related to productive and institutional factors of certain geographical areas. These advantages are a consequence of the exploitation of location advantages, such as, cheap labor, raw materials, and smaller transportation costs, among others. Locational advantages are those which are specific to a country.

Thirdly, internalization advantages (I) derive from the firm's capacity to manage and coordinate activities throughout its value chain thus, generating more added value than its competitors. These advantages are related with the integration of transactions inside multinational hierarchies through foreign direct investment. In short, internalization advantages are the benefits that derive from internal markets and that allow firms to bypass external markets and the costs associated with them.

This model is an attempt to integrate existing theories into one universal model. It exposes the "why", "where" and "how" of a firm's internationalization. If advantages do exist, the company should explore its assets through international production, as opposed to exporting, joint ventures, licensing or franchising.

As mentioned above, the eclectic paradigm is a synthesis of other approaches that concentrate on trade or international production, possession of superior technology or imperfect market structures. Accordingly, there is neither swift perspective on the competitive nature of international production, nor any consideration of collusion and/or market power.

### 3.3 The Uppsala theory

The Uppsala internationalization model suggests that internationalization is sparked by an evolutionary and sequential commitment through time. Johanson & Vahlne (1977) developed this process model, which proposes a sequential learning step-by-step process, based on the study of four Swedish companies. This model proposes four sequential phases that represent higher degrees of international commitment:

1.  No regular activities or indirect exports;
2.  Direct exports through agents or independent representatives;
3.  Direct exportation through own subsidiaries abroad;
4.  Foreign direct investment through the establishment of production or commercialization units abroad.

The firm initiates its process with non-regular export activities to neighboring countries or countries with low psychic distance, therefore, avoiding uncertainties and high risk-taking. In this phase firms rely on foreign direct investment as they lack relevant information about the market. In the second phase, the company begins selling its products abroad through independent representatives, creating a channel of information between the company and its foreign market. Thirdly, with the establishment of an international subsidiary, the firm tries to create its own information channel in order to obtain a larger control of operations abroad. Finally, the establishment of assembly plants and subsidiaries abroad give companies control over their production and sales (Morgan & Katsikeas, 1997).

It is important to note that these phases are not mandatory and that knowledge plays an important role in this evolutionary process. Depending on the way that they gain experience during their interaction in foreign markets, firms do not always follow all the phases.

This model is based on the concept of psychological distance. In this way, firms enter foreign markets where and when they spot opportunities and look to diminish the uncertainties they face in their international expansion. Accordingly, firms gradually extend their activities to new, more distant markets from the psychological point of view (Hansson et al., 2004).

This model suggests that the absence of knowledge about foreign markets is the main barrier for a larger international commitment; this obstacle can only be overtaken by experiential learning.

## 3.4 The internationalization network theory

Based on the Uppsala model, Johanson & Vahlne (1990) continued their study on internationalization, applying the network perspective to their theory. The extension of the model involves investments in networks that are new to the company, seeing that international penetration means developing positions abroad and increasing resource commitments in the network in which the company has developed its positions. The integration of this network can be understood as the coordination of different national networks the company is involved with (Ruzzier et al., 2006). Therefore, the relationships among firms are seen as networks; firms then internationalize because other firms of the (inter)national network do. Within the industrial system, firms involved in the production, distribution and use of goods and services are mutually dependent on their specialization patterns. Given the configuration of the world economy, certain industries or kinds of markets are more inclined to be internationalized (Buckley & Ghauri, 1993; Andersen, 1993).

Johanson & Mattsson (1988) suggest that company success upon entering new international markets is more dependent on its position in the network and its relationships in present markets, than in the cultural characteristics of international markets. Johanson & Vahlne (1992) demonstrated that entering foreign markets is the result of a gradual interaction process among the parties and, of the development and maintenance of the relationships throughout time. The network perspective goes even further than incremental internationalization models, suggesting that the firm strategy arises as a behavioral pattern influenced by the range of relationships within the network (Rundh, 2003).

The network approach, including the micro and macro perspective defined by Johansson & Mattson (1992), considers that networks are constituted by the businesses that firms maintain with their clients, distributors, competitors and governments. Johanson & Mattsson (1992) argue that, as firms internationalize, the number of actors and interactions increase and, as a consequence, the relations among them become narrower.

During the internationalization process, firms create and develop commercial relations with their counterparts abroad. This process evolves in different ways: first, creating relations with partners in countries new to the firm – international expansion –, secondly, enhancing the commitment in the networks the firm is already involved in – international penetration – and finally, connecting the existing networks in different markets – international integration (Hansson et al, 2004).

The strength of the internationalization network model lies in the explanation of the process and not in the existence of multinational or international firms. From the network point of view, the firm internationalization strategy can be characterized by: the need to minimize the development of new knowledge; the need to minimize the adjustment to new realities; and the need to explore the firm's positions in established networks (Johanson & Mattsson, 1992).

In this way, network activities allow the company to maintain relationships and to underpin the access to new resources and markets. The network in which the firm is most active is the main engine of the firm's internationalization process. This was not the reality in the Uppsala model, which contended that the internationalization process depended on firm specific advantages or on the psychic distance of the market (Hansson et al, 2004).

The main contribution of the network theory lies in recognizing that a firm's internationalization process is not the effort of a single company, but the result of the relationships among many firms. Table 1 summarizes the main concepts of the different theories analyzed.

| Theory | Main Authors | Key concepts |
|--------|-------------|-------------|
| Internalization theory | Buckley & Casson | • The choice of the transaction mode varies according to specific cost;<br>• Internalization of activities the market performs expensively or poorly, through vertical integration;<br>• Larger firm control;<br>• Monopolization of knowledge. |
| Eclectic Paradigm | Dunning | • The internationalization of firms is explained by three factors:<br>1. *Ownership* – The firms invests in a foreign market using its core competences as a competitive weapon vis-à-vis its main competitors.<br>2. *Location* – The firm internationalizes choosing the market with the best conditions for the firm.<br>3. *Internalization* – The firm invests in facilities abroad in order to internalize the operations that were performed by the market. |
| Uppsala Theory | Johansson & Vahlne | • Internationalization is seen as a process that integrates a gradual, continuous evolution through which firms acquire experience and knowledge by progressing in their involvement and commitment in foreign markets. |
| Network Theory | Johansson & Mattsson | • The internationalization of a firm is the result of the development of a network of (internal and external) contacts with individuals or firms that possess resources and knowledge/experience, in which the access to information and knowledge is more accessible and less costly. |

Table 1. Synthesis of contributions of the internationalization theories covered

## 4. Historical evolution of the automotive industry in Portugal

### 4.1 The special EDFOR case (1937-1952)

The first milestone of the Portuguese auto industry took place in the first decades of the XX century, as in the rest Europe. In 1937, Eduardo Ferreirinha and Manoel de Oliveira – who were fans of car races in Portugal – invested in the foundry of car components. As a result of this passion, they managed to develop a sporting vehicle, with a Ford V-8 engine, the body and skeleton melted in an aluminum league, with a driver-control suspension mechanism, named EDFOR. Its production included several foreign components, foreseeing the current platform concept (Féria, 1999). Although the first steps were quite promissing for the national auto industry, the mass production of the EDFOR prototype never took place due to the beginning of World War II.

## 4.2 Automotive assembly lines (from 1950s to 1960s)

In the beginning of the 1950s, Portugal lived in a closed economy. The protectionist mentality dramatically influenced the Portuguese industry, which was protected from competition by strict customs regulations, by *industrial conditioning*[1] and by cheap labor that reflected negative growth year after year (Féria, 1999).

Following this protectionist mentality, the Portuguese government adopted strict measures, similar to those that occurred in several developing countries. Such measures imposed import quotas and closely followed the policies that had been implemented in Spain in order to achieve a national car brand. These stringent policies affected mainly passenger vehicle manufacturers that did not assemble their products on national territory. By this time, the development of the Portuguese auto industry was quite widespread as the producers of the main commercialized brands present in Portugal were forced to introduce their assembly lines in order not to loose market share (Féria, 1999).

Several investments took place in Portugal at that time: among other, factories from GM/Opel, Ford, Citroën, Fiat, Barreiros and Berliet. Investment in heavy vehicles also took place. Although the auto industry investment was thriving, the market was far from working properly, due to the abovementioned reasons as well as strict regulations imposed by the government. Likewise, as referred by Féria, (1999: 11) the assemblers, with one exception, "never invested on improving the local supply chain, barely investing in the promotion of the national components industry".

The greatest consequence of this policy was the high final price of the vehicles. This was the result of added production costs based on insufficient critical mass and lack of production competences, which hindered productivity. During this era, the Portuguese auto industry experienced several successful cases – the Ford P-100 is a clear example –, with pervasive consequences in terms of exports until 1974-75. This success managed to deploy the creation of industrial jobs, although with poor levels of qualification (Féria, 1999).

During this era, as a result of the policy implemented, several assembly lines were created. One of them is still in operation – in Ovar – assembling several commercial Japanese vehicles for Toyota.

In parallel, as a consequence of several investments in the auto industry, the production of components – glass, upholstery, car seats and other interior components – took place in order to supply auto assemblers.

Afterward, this police revealed itself as inadequate to the growth of the auto industry in Portugal, following other unsuccessful attempts to produce "made in Portugal" vehicles.

---

[1] During the *industrial conditioning era,* the government only authorized the creation of new firms if they did not jeopardize the economic behavior of firms already competing in the Portuguese economy.

It is important to notice that this period, although based on the stringent policy implemented during the 1950s, underpinned an industrial change that managed to transform the national automotive industry.

### 4.2.1 FAP – Fábrica de Automóveis Portuguesa (1959-1965)

Towards the end of the 1950s, in parallel with the frustrated attempt to encourage the builders of the most traded brands in Portugal to install their assembly plants in the country – in order to stimulate the development of the national industry, a nationalistic initiative – "the men of the establishment – takes place. Although they had little knowledge about the automotive industry and all its industrial complexity, they were enthusiastic about the production and commercialization of an economic family vehicle, produced under the license of an assembler. They set up a new factory – Fábrica de Automóveis Portugueses (FAP) – and after the initial investments in land and infrastructure, they initiated their research for potential licensors interested in their vehicle (Féria, 1999). Soon after, they realized the complexity of such achievement – due to a narrow market segment, lack of industrial tradition and very unlikely international penetration of their product – and that their project was doomed, given to clear economic inferiority of the feasibility of the factory.

In 1963, there was a strategic change and FAP started to plan the production of agricultural tractors, substituting the original idea of producing commercial vehicles. Concentrating their effort in this new direction, FAP obtained a licensor for assembling tractors, investing in training the employees and bargaining state subsidies and financing. However, FAP rescinded its functions without ever producing a single vehicle or having started scale production of tractors (Féria, 1999).

This landmark, although apparently of little relevance for the Portuguese car industry, reveals the importance of understanding the complexity of the car industry as part of a broader, global industry.

### 4.3 Protected market (1961-1974/76)

### 4.3.1 Public industrial policy

In 1962, imports substitution marked the political orientation of the automotive industry (Felizardo et al., 2003). The "Assembly Law" influenced industry dynamics, which were characterized by strong state intervention. This law imposed the assembly – in Portugal – of all vehicles commercialized in the domestic market as well as, the restraining of imports to a maximum of 75 units per manufacturer of Completely Built Up (CBU) units. This law also imposed that the national added value increase should be at least 25% for the units assembled locally (INTELI, 2003). Given the focus on the supply of the domestic market through domestic production, the liberalization of imports of Completely Knocked Down (CKD) units was kept. According to these impositions, international manufacturers were allowed to import merely 75 CBU units per year and unlimited CKD units. However, the national manual labor incorporation rate had to be at least 15% of the cost of the complete unit (Felizardo et al., 2003).

Given the stringent limitations imposed by the legal framework, car manufacturers were ruled by a specific and restricted program of national incorporation. On the other hand, one step back in upstream activities in the value chain, components producers were receiving indirect incentives to develop subsidiaries according to the degree of the national incorporation of their activities (INTELI, 2003). This was an indirect way of supporting the national automotive industry.

This legislative setting, which represented the first sectoral legislation in the automotive industry in Portugal, was kept until 1972. However, there were some slight changes, such as the authorization to import more than 75 CBU units for manufacturers from EFTA member countries in 1968 and, the increase of the rate of mandatory national incorporation to 25% of the value of the complete unit in 1969 (Felizardo et al., 2003). Afterwards the sectoral legislation was altered as a consequence of ineffective productivity of assembly lines and of the commitments the national government intended to fulfill in the international arena.

The "Assembly Law" constituted a milestone in the automotive industry sectoral policy during the 1960s and 1970s, introducing in Portugal, for the first time, a model of industrialization in the assembly of vehicles (Felizardo et al., 2003). Throughout time, the national legislation changed to "serve" the automotive industry as well as the car components industry, according to the evolution of the sectoral fabric.

The external recession – based on the first oil crisis – and the national social and political revolution – originating from the Revolution of the 25 of April of 1974 – did not bring about a new sectoral policy or legislative setting (Felizardo et al., 2003). Although legislation during this turbulent period was erratic, unstable and incoherent, only in the next decade did new sectoral law emerge.

### 4.3.2 Assemblers strategic line

As a consequence of the sectoral policy, which imposed limits on imports to the Portuguese market, there was a proliferation of assembly units due to foreign direct investment and licensing contracts. In an attempt to face the legal framework imposed, the biggest car manufacturers decided to invest in Portugal, opening new factories on Portuguese territory in order not to loose their market. The majority of the assemblers present in Portugal granted assembly licenses to importers or other national companies. BMC, Citroën, Ford, GM, Renault and Fiat were the six multinational subsidiaries of the auto industry that initiated production, especially of commercial units, in Portugal (INTELI, 2003). This behavior had a pervasive effect on the diffusion of assembly units, on the number of traded brands and, on the range of models produced, driving the expansion of the national output of vehicles and the growth of the market, as presented in Table 2.

Through time, the number of companies operating in Portugal began to increase substantially – as well as the number of produced units, reaching a peak of 101,406 units in 1974 and clearly reflecting the limited size of assembly lines operating in the Portuguese industry (Felizardo et al., 2003).

This industrial setting led to a fragmented demand based on low scale volumes, low-technology and low-complex piece parts in which the economic added value was predominantly based on cheap manual labor. This situation did not contribute to the improvement of the endogenous characteristics of Portuguese automotive industry. Only during the mid-1970s could indigenous incorporation be measured in function of units produced in Portugal. However, and due to the very limited number of vehicles produced

| Year | Number of assembling units |
|------|----------------------------|
| 1962 | 2 |
| 1964 | 17 |
| 1970 | 18 |
| 1974 | 21 |
| 1976 | 20 |

Source: Selada and Felizardo (2002)

Table 2. Number of assembling units: 1962-1976

on each assembly line, the production of piece parts was inefficiently based on limited production volumes, even when considering the aftermarket indicators (INTELI, 2003).

The assembly of vehicles in Portugal was destined basically to the domestic market, in the segment of light vehicles – until the end of the 1970s – and commercial vehicles – until the end of the 1980s. The strategies of assembly units were particularly focused on commercial objectives, assembling a wide range of brands and models with industrial units characterized by producing – intermittently and inefficiently from the economic point of view, due to the lack of economies of scale – small series. However, despite all those adverse factors, at the end of the decade, the automotive industry was responsible for approximately 25,000 jobs, including assembly lines and their suppliers (INTELI, 2003).

### 4.3.3 Strategic lines of auto components

During the protected market era, the majority of assemblers in the Portuguese auto industry – with the few exceptions of GM/Opel, Ford and Citroën – did not possess a deep knowledge on how to improve their relationships throughout the value chain, relegating the components industry to a secondary place in industrial policy. This situation was aggravated by the absence of direct mechanisms to support to the development of the components industry, such as legal framework regarding the incorporation of indigenous piece parts. On the other hand, the poor labor productivity, the lack of modern industrial capital equipment and, the disposal of valuable material were inhibiting factors for the assemblers to source their components from the local car components industry. Parallel to this situation, the legislation at that time did not allow auto assemblers to produce components, which indirectly imposed to assemblers a clear dependence on their supplying subsidiaries (Felizardo et al., 2003).

By the end of 1970s, 170 components manufacturers operated in Portugal, supplying the existing assembly lines. However, they were not exclusively focused on the auto industry, complementing the production of components and piece parts with the development of other businesses. The auto components suppliers employed around 15,000 people (Felizardo et al., 2003).

The auto components industry was characterized by a myriad of small artisanal, inefficient firms, with low levels of quality and limited organizational, commercial and technological competences. The industry was exclusively oriented to the domestic market and focused on manufacturing traditional technology parts – metallic pieces, batteries, glasses, tires, seats and other nonmetallic parts – of low added value. Due to the protectionism of the national market, production was limited and intermittent, which inhibited investment and specialization. Moreover, and as a consequence, the prevailing technologies were rudimentary characteristics emphasizing simple production processes (Felizardo et al., 2003).

In parallel, the auto components industry did not attract foreign direct investment until 1979 (Felizardo et al., 2003).

### 4.3.4 Alfa-Sud project (1972)

In the end of the 1970s, Alfa Romeo was owned by a prestigious manufacturer producing several car models with luxurious and sporting characteristics, targeted to the European high purchasing power elites.

In order to expand its activities to other market segments – in particular younger public and middle and upper social classes –, Alfa Romeo developed, during the last years of the 1960s a concept car: Alfa Sud. This concept was designed with Alfa Romeo's traditional sporting lines and with less opulent aesthetics. The objective was to place Alfa Sud in the European market, with very a competitive final price thus, crushing main competitors in the small family vehicles segments (Féria, 1999).

In order to achieve this objective – and in an attempt to increase its competitiveness and lower the unit costs of production facilities – Alfa Romeo planned to delocalize Alfa Sud's production. Portugal was part of the selection short list of potential locations due to the cheap cost of labor (Féria, 1999).

In 1972, Portugal was living the first years of a more liberalist government, opening a window of promising social, political and economic opportunities. The industrial policy was trying to unleash the power of foreign direct investment and cool down the protectionist vein of the *industrial conditioning era*. Accordingly, the government received the Alfa Romeo's proposal enthusiastically and realistically, creating, at once, a multidisciplinary team to analyze the national capabilities of supplying fundamental key components to the project. In fact, for the very first time, a structuralist approach was addressed taking into account the whole value chain and not a situational perspective towards a foreign direct investment possibility.

Several characteristics – namely those related to the supply chain of the Alfa Sud – were analyzed as determinant factors for the project to be successful: organizational and managerial styles, technology and existing production equipment and, the existence of quality processes and products. Given the absence of these key characteristics in the Portuguese industrial arena, mainly those related to the quality standards required by Alfa Romeo, the Alfa Sud project was not located in Portugal. Although, there were several firms supplying auto assemblers with very high quality products, Portugal, as a location, was senseless as it would force Alfa Sud to import almost all the components, creating unusually complex logistics at the time (Féria, 1999).

The Alpha Sud experience was, according to Féria (1999), very insightful as it gave Portuguese industrial planners the opportunity to analyze the auto components suppliers' capacities, knowledge and technological potential. Accordingly, this attempt, although not a prominent milestone in the history of the Portuguese auto industry, constituted an insightful and fruitful opportunity with important repercussions for future projects – to address the auto industry from a systemic perspective.

## 4.4 Sines Flop (1979-82)

In 1979, the Ford Motor Co. European Board analyzed the possibility developing a new project in the Iberian Peninsula. In search of increased competitiveness and based on the aftermath of a successful investment carried out in Valencia, Spain, the board was looking for geographical areas to underpin Ford's delocalization strategy, and decrease production costs. The board entrusted the site selection team with the responsibility of finding a location for the production of the new *mini Extra* (Féria, 1999).

Portugal came out as one of the potential locations, since one of main Portuguese policy objectives was to avoid loosing another manufacturing implementation to neighboring Spain. Following the contacts of the selection team, national industrial authorities attributed maximum priority to Sines on the South of Portugal due to the strong investment carried out in the Sines industrial park. Ford intended to increase its foothold in the European market and Portugal would perform an important role in this expansion. However, the site selection team cast some doubts about the Sines location due to the lack of trained labor and technology endowed suppliers. Even though negotiations were kept, in 1982 the final negative decision was released: Ford was not committed to invest due to the lack of technologically driven supply and the changing European market conditions. Eventually, this decision led to wide scale success the *Fiesta* – the main vehicle produced in Valencia, Spain. Ford definitely abandoned the *Extra* project, collapsing the Sines automotive industrial dream (Féria, 1999).

As in previous failures, this experience allowed the Portuguese authorities to have contact with the reality of the automotive industry, namely, with the inherent demands of this sector.

## 4.5 The renault project (1977-1986/88)

### 4.5.1 Public industrial policy

During this period, the automotive sectoral policy orientation was marked by the promotion of exports, due to a specification of a production quota to all assembled CKD units with less than 2000kg. The legal framework created barriers to auto assemblers and all new investments were subject to applications. However, there were some exceptions: those in which the State was the largest shareholder. Subsequent legislation widened the range – to CKD and CBU units – of exportable products and an additional contingent of CKD products could be obtained by exporting products manufactured in the same manufacturing unit. The promotion of exports was also encouraged for other Portuguese products that were destined to vehicle assemblers and to the auto components manufacturers produced by Portuguese firms and supplied to exporting assemblers (Felizardo et al., 2003).

This new regulation, destined to the auto industry, was complemented with incentives to foreign direct investors – based on quality of human resources, cheap labor and geographical location – in order to create the endogenous condition for the automotive industry to flourish (INTELI, 2003).

State interventionism led to the creation of the Cabinet for the Automotive Industry Studies (1974) and to the Automotive Industry Commission (1976). With this deep change in the Portuguese auto industry, in 1977 the Portuguese authorities invited several international car manufacturers to invest in manufacturing facilities in Portugal (Felizardo et al., 2003). Renault and to Citroën were among the short list of candidates. Renault was chosen and in February of 1980 the foreign investment deal was signed with the *Régie Nationale des Usines Renault* (RNUR). According to Féria (1999), this was a politically-based decision. According to Felizardo et al. (2003), Citroën's proposal was excluded because of low national components incorporation.

Before the Renault project started, the government published new legislation for the automotive industry (Felizardo et al., 2003) restraining the import of CBU vehicles and establishing quotas for the incorporation of national components in CKD vehicles. Clearly, industrial policy's intention was to create a critical mass of assemblers, to create jobs and to underpin the technological development of the auto components industry. From the international business perspective, the main objective was to abandon the import substitution policy Portuguese governments had been using and to embark on an export promotion policy.

New incentives were established for the industrial conversion of assembly lines as most of them lacked technological conditions to adequately compete in the market. Initially, the incentives were targeted to car manufacturers producing units with less than 2000 kg and the conditions were very simple: those implementing the change would be granted the possibility to transfer their import quotas – CBU units – to their production – CKD units – and export quotas. Afterwards, commercial units were also included (Felizardo et al., 2003). In summary, inefficient assembly lines and components producing units were shut down or reconverted. At the same time some new firms were created – in order to meet the European market dimension (INTELI, 2003).

After the conversations with the European Economic Community (EEC) authorities – with Portugal seeking to enter the EEC – the end of the protection was stalled and a new protocol (31/12/1984) was negotiated that basically ratified the principles and the underlying mechanisms of the legal framework at the time (Felizardo et al., 2003).

On the other hand, the corporate institutionalization of components manufacturers was more difficult and less powerful than that of auto assemblers of the auto industry. Hence, AFIA (Association of Automotive Industry Manufacturers) was created in 1979 (Felizardo et al., 2003).

The legal framework ceased its validity in 1984. During the same year law n° 405/84 was set up aiming to regulate the auto industry until the definite entrance of Portugal into the EEC. Portuguese protectionism ended in December of 1987. Although between 1984 and 1987 there was a change in the state of mind of all players in the Portuguese automotive industry, 1988 brought about the complete liberalization of the car market (Felizardo et al., 2003).

### 4.5.2 Assemblers strategic line

The exports promotion policy represented an opportunity to reconfigure inefficient assembly lines. As a consequence, Portugal witnessed a shrinking number of manufacturing units as well as a decrease in the quantity of models and brands manufactured in Portugal. The closure of assembly lines affected: a) the licensed units more than the branches of multinational units, and b) the companies producing passenger cars more than those producing commercial vehicles.

While between 1979 and 1982 there were 19 assemblers operating in Portugal, in 1984 the number of assemblers decreased to 16, in 1986 to 13 and in 1988 only 12 units were in operation. According to Table 3, between 1979 and 1988 ten companies either closed down or reconverted their assembly lines.

| Assemblers | Closure date |
|---|---|
| A.C.P. de Motores e Camiões | 1979 |
| Garrido e Filho | 1983 |
| Imperex | 1983 |
| IMA | 1984 |
| Comotor | 1984 |
| Montavia | 1984 |
| Somave | 1986 |
| Entreposto | 1986 |
| Proval | 1987 |
| UTIC | 1988 |

Source: based on Selada and Felizardo (2002)

Table 3. Firms that closed down their facilities

The change of mind in industrial policy drove some firms to reinforce their presence in Portugal; namely GM, founded in 1963. GM focused its strategy on exporting components to other European GM units, which gave the GM local unit an unprecedented scale. In order to increase its national market share, GM decided to import CBU vehicles, increasing the added value and the relationship with Portuguese suppliers (Felizardo et al., 2003).

The Renault project was a major turnaround as Renault invested, for the first time in Portugal, in the creation of a complex, modern infrastructure. This project was one of the most important in the Portuguese industry, whose investment was composed of three units:

- The setup of a car manufacturing unit in Setúbal, with an output capacity of 80,000 vehicles/year and with a level of national incorporation between 50% and 60% of this output;
- The creation of a mechanical components factory, in Cacia, with an output capacity of 80,000 gearbox/year and 220,000 engines/year, with a national incorporation level between 60% and 80% of the output;

- The turnaround of the Guarda factory, in order to shape this industrial unit to an export oriented unit;
- The setup of a foundry unit – FUNFRAP – to supply the engines and gearboxes unit.

In 1980, Renault started the production of its R5 model in Setubal and, in 1981, they initiated production of car components in Cacia. In the meantime, after some change processes, the Guarda factory was subsequently sold. The foundry unit started its activity in 1985, producing melted parts for engines (Felizardo et al., 2003).

According to Table 4, the factory went through a significant increase in its output since the beginning of its activity. The level of output components, which initially was based on gearboxes and subsequently widened to engines and mechanical components, is displayed in Table 5.

| Year | Output of Renault Factory (units) |
|------|-----------------------------------|
| 1980 | 3.006 |
| 1981 | 27.895 |
| 1985 | 28.123 |
| 1988 | 44.475 |

Source: based on Felizardo et al. (2003)

Table 4. Output of Setubal Renault's factory

| Year | Production of components | |
|------|------------------------|---------|
|      | Gear boxes | Engines |
| 1981 | 6.699 | - |
| 1982 | 53.525 | 47.787 |
| 1988 | 82.695 | 226.885 |

Source: based on Felizardo et al. (2003)

Table 5. Renault's components production units

During the 1980s, the Portuguese auto industry changed extensively after an initial reduction of the number of vehicles produced, reaching 75,675 units in 1979. Due to an economic crisis, the beginning of Renault's operation meant that the number of cars manufactured in Portugal reached a peak in 1982: 118,958 units. However, between 1982 and 1986 the output diminished again to 96,006 units. Even with the evolution of the Renault project, national and international conditions limited the development of the automotive industry in the first half of the 1980s. From 1986 to 1988, the auto industry witnessed the recovery of output production, which reached 136,524 units, in 1988, being most of them for the foreign market (Felizardo et al., 2003)

### 4.5.3 Strategic line of auto components suppliers

One of the main objectives of industrial policy was the promotion of exports, which was achieved due to a heavy investment (and openness) in the auto components industry. The possibility of compensation in the exporting of components underpinned the creation of new firms and jobs, which created new industrial dynamics.

The Renault project sparked the development of the auto components industry of in Portugal, as it paved the way for high levels of incorporation of Portuguese manufactured components and the establishment of some companies associated to the French manufacturer.

The positive effects related to the Renault project that influenced the development of the Portuguese components industry are the following:

• Suppliers quality certification;
• Product certification and homologation;
• The introduction of modern production management processes;
• Human resource qualification;
• The introduction of new marketing, organizational and technological learning processes;
• The introduction of the auto industry *modus operandi*;
• The promotion of exports and contacts with the global auto industry;
• The promotion of foreign direct investment in Portugal.

The components industry evolved positively for a long time based on the external market, which in 1986 overtook the domestic market (Felizardo et al., 2003).

Of equal importance, the Portuguese auto industry could rely on several domestic components suppliers embedded in the "auto industry culture". These were concerned with delivering products that competed in terms of quality, costs and time delivery. The firms had the capability of widening their product range. Furthermore, they could produce technologically complex products due to their investments in emerging technologies as well as in process innovation technologies. The knowledge generated by their relationship with more technologically oriented clients also contributed to the dissemination of the of specialization processes and to the adaptation of a new and demanding environment. Some components suppliers that managed to improve their performance in the domestic auto industry were invited to internationalize their activities and, consequently, had the chance to acquire the first contacts with the global auto industry. Some of them managed to progress from transactional relationships to relational partners (Moreira, 2005).

From the beginning of the 1980s, Portugal witnessed a strong increase of foreign firms entering the auto components industry, providing their suppliers with an excellent opportunity to have a foothold in the Portuguese market as well. Due to the international economic turmoil, the number of auto components firms diminished though their production output increased (INTELI, 2003).

In the end of 1980s, Portugal had a competitive fabric of suppliers with a strong set of production process competences, having cheap labor as its main asset (Felizardo et al., 2003). In the following years, the national firms' investments and turnarounds started to show very good results due to the exports boom in 1985 (INTELI, 2003).

In 1988, Portugal completely opened its markets to products from the EEC and a new phase of development in the Portuguese automotive industry took place. In spite of the imports growth, the exports of vehicles and of the production of components continued to grow (INTELI, 2003). In the same year, the first Specific Program for the Development of the Portuguese Industry (PEDIP) was launched in an attempt to diminish the gap between Portugal and the developed European countries. The program involved several actions in a wide range of areas, such as R&D, international business and financial support to all firms competing in the market. In addition, it offered incentives for the creation of new foreign companies (INTELI, 2003).

For the Portuguese Government, the establishment of an original equipment manufacturer would be a unique opportunity for the development of the Portuguese auto industry as well as for the auto components industry (INTELI, 2003).

## 4.6 The UMM project (1977-1993)

In 1977, a new firm was created in Lisbon – UMM. This firm's main objective was to produce and commercialize all-terrain vehicles. In fact, UMM developed niche vehicles (Féria, 1999) – CPE (model with "narrow door"), CPL (model with "wide door"), ALTER I and ALTER II – and its main clients were the Portuguese army, the Fiscal Guard, the police – Republican National Guard (GNR) – and Electricity of Portugal (EDP), one of the main Portuguese target audience at the time. Nevertheless, the remaining branches of the army did not purchase the UMM vehicles, even after strong technical improvements were implemented. As a consequence, the UMM output never reached the scale it could have achieved if the public fleet could have been sourced by UMM (Féria, 1999).

According to Féria (1999), in spite of its technical problems and its failure, this project could have reached an interesting development if the Portuguese authorities had invested more thoughtfully in R&D activities (the technical problems should not have hindered UMM's market pervasiveness).

This landmark of the Portuguese auto industry reflects a bipolar culture: on the one side, a nationalist culture focused on the development of a national brand – with plenty of supporters, at the time – and, on the other hand, a pragmatist culture in which the project did not succeed due to the lack of strong technical competences. In fact, with Portugal's entry in the EEC and its free market approach, international competitors ended the "national brand" dream.

## 4.7 The golden period of foreign investment (1987)

After a long bitter period of in which large multinational firms ignored Portugal (due to the revolutionary period and the dismantling of international players) in 1987 foreign direct investment (FDI) began to occur.

Several investments took place. Ford Motor Co. Electronic Division, Continental, Delco-Remy (GM), Samsung, COFAP (Brazil), to PEPSICO (U.S.A.) are some examples of FDI that the Portuguese government managed to persuade with several types of subsidies. These projects, directly or indirectly, were related to the auto industry. Their approval and development paved the way for the Autoeuropa project, a joint-venture between Ford and Volkswagen (Féria, 1999).

## 4.8 Autoeuropa project (1989-2011)

### 4.8.1 Public industrial policy

In this phase, the auto industry policy was marked by the reopening of the market. The liberalization of the market took place in 1988, after a 25-year period of protectionism. With this new legal window, the importing of vehicles from the EEC, EFTA and other preferential countries (Yugoslavia, Cyprus, Malta and Lomé Convention countries) was liberalized. Furthermore, new restraints were created to import vehicles from other countries such as Japan, South Korea, U.S.A. and Brazil, despite total freedom for importing CKD vehicles from those countries (Felizardo et al., 2003).

The strong increase of foreign direct investment reflected State intervention – based on heavy investments in technological development and innovation – that attributed direct subsidies and conceded fiscal emoluments. At the heart of this industrial policy, PEDIP, PEDIP II and POE programs were created to grant incentives for the companies investing in developing their competences – technology, innovation, quality, training, management, marketing, among others.

In 1989 Portugal learned of Ford Motor Co. and Volkswagen's intention to establish a new joint manufacturing unit in Palmela, Setúbal (Féria, 1999). After a long period of negotiations, the investment and incentives contracts were signed in July of 1991; the launch of the Autoeuropa project took place (Felizardo et al., 2003).

The National Institute Supporting Small and Medium-sized Firms and Investment (IAPMEI) created a Cabinet (GAPIN) in order to stimulate and develop the supply potential of the auto components suppliers. The Autoeuropa project was used to diffuse joint-ventures opportunities among foreign and national firms and to promote the development of new competences and capabilities. It was also used to improve product quality in upstream activities throughout the supply chain (Féria, 1999).

In April of 1995, Autoeuropa was inaugurated. Four years later, the end of the joint-venture between Ford and Volkswagen was announced. Autoeuropa took Ford's position and continued production of Ford's multi-purpose vehicles (MPV) until the end of 2004. When the agreement with the Portuguese State finished, industrial incentives accounted for approximately 490 million euros. A new agreement was signed and the State committed to another 12 million Euros of incentives. Ford Galaxy's MPV ceased its production in February of 2006. Autoeuropa now produces the EOS and Scirocco (Felizardo et al., 2003).

The Center for Excellence and Automotive Industry Innovation (CEIIA) was created in 1999 in order to promote networking activities between all stakeholders of the auto industry (Felizardo et al., 2003).

### 4.8.2 Auto manufacturers strategic axes

With the reopening of the market, the rationalization and conversion of assembly lines was intensified due to the limited market size and to the trade liberalization in EEC countries.

There were 10 assembly lines operating in Portugal 10. This figure decreased to 8, in 1994, and to 7 in 1997. Since 2002 until 2004 only five assemblers operated in Portugal:

Autoeuropa (Palmela), PSA Peugeot Citroën (Mangualde), Mitsubishi Spindle Trucks Europe, Opel Portugal (Tramagal) and Toyota Caetano Portugal (Ovar). As shown in table 6, those firms that operated under license agreements closed down their facilities and several multinational subsidiaries decided to assemble only commercial vehicles, less demanding from the technical point of view and with less economies of scale.

In what concerns to the closing of industrial units, the Renault factory, in Setubal, assumes special importance by its structural impact on the Portuguese auto industry, as seen above. The slow demand growth of international markets and the possibility to supply the market from other European factories (producing the Clio) forced Renault, from 1992, to slow down its output in Portugal, leading to a significant reduction of employees. In fact, Renault's strategic interests in Eastern European markets as well as its factory located in the Slovenia, discouraged RNUR to continue its involvement in the Setubal's factory. Social and economic difficulties led Renault to close down its manufacturing facility in Setubal in 1995 (Felizardo et al., 2003).

| Manufacturing units | Closure date |
|---------------------|--------------|
| Reicaab             | 1991         |
| Soma                | 1992         |
| Movauto             | 1993         |
| Baptista Russo      | 1995         |
| Movar               | 1995         |
| Renault/Sodia       | 1998         |
| Ford Lusitana       | 1999         |

Source: based on Felizardo et al. (2003)

Table 6. Factories that closed down: 1991-1999

The Autoeuropa project – with the aim to produce three brands: Ford Galaxy, VW Sharan and Seat Alhambra – came out as the new engine of the Portuguese auto industry, with a production capacity of 180,000 vehicles/year. This project attracted at once 22 new foreign units; eleven of them set up their facilities nearby Autoeuropa's industrial park, in order to be able to implement just-in-time methodologies (INTELI, 2003). The 225 million Euros of investment enabled the creation of 5,200 direct jobs and between 7,000 to 10,000 indirect jobs. According to Felizardo et al. (2003) the national incorporation reached the 45%.

The Autoeuropa's production reached its peak in 1998 – 138,890 units – and has been decreasing every year since. Table 7 presents some Autoeuropa's indicators for the year 2008.

### 4.8.3 Strategic lines of auto components suppliers

The opening of the Portuguese economy, the accession to the EEC and the Autoeuropa's investment in Portugal brought about quantitative as well as qualitative changes to the

| Indicators | 2008 |
|---|---|
| Number of units produced | 93,609 |
| Sales (Millions of Euros) | 309.4 |
| % Portuguese Exports | 10% |
| % PIB | 1% |
| Number of workers | 3,028 |

Source: Volkswagen Autoeuropa (2008)

Table 7. Autoeuropa's indicators and its impact on the Portuguese economy

Portuguese auto components industry. Autoeuropa had an extremely important role in the consolidation of the car components industry as it represented a window of opportunities towards international markets to several Portuguese suppliers. The specialization pattern of the Portuguese industrial structure changed not only as a consequence of the investments of multinational firms, but also because the Portuguese auto industry had the opportunity to conquer new markets, though suffering the consequences of international competition.

According to Felizardo et al. (2003), the presence of Autoeuropa's project paved the way to a deeper involvement between the international fabric of Autoeuropa's network and the indigenous auto suppliers – in 2008, they were more than 60 –, leading them to competitive advantages that otherwise would not have acquired.

The levels technological demand imposed by Autoeuropa impelled a positive reaction from Portuguese suppliers, which was the result of consolidated knowledge acquired with the Renault experience. Throughout time the number of Q1 certification increased; today is a basic, fundamental pillar for competing in the market.

Autoeuropa, in the heart of its network of suppliers, created a lean production environment at several levels, managing to transfer technological, organizational, relational and managerial know-how to Portuguese suppliers. Moreover, the creation of joint-ventures between foreign firms and indigenous suppliers created conditions for some of them not only to supply components to Autoeuropa, but also to internalize know-how and technical support, which opened the possibility of exporting to new markets and to be integrated in international networks (Moreira, 2005).

Aueuropa's investment dynamics generated a new wave of investments in Palmela, with positive impact on the auto industry as well as on the Portuguese economy.

As a consequence of Autoeuropa's investment in Portugal, between 1989 and 2001, the components industry evolved positively increasing its sales volume in 3,229 million Euros, representing a growth on exportations from 584 to 2,642 million Euros. As a consequence, the components industry started to play one of the main roles in the Portuguese economy, side by side with textile and the clothing industries. The components industry evolved towards an integrated and consolidated network of firms, involving indigenous and foreign firms supplying assemblers in Portugal as well as abroad (Felizardo et al., 2003).

The main products manufactured by the sector are the following ones: engine components, transmissions, brakes and electronic components (Felizardo et al., 2003). In the components sector there is a predominance of metal-mechanic firms, metal stamping firms, plastic components firms and electronics components firms. There was a strong investment in process innovation technologies as well as in the control and organizational integration, within and between firms, led by lean production Autoeuropa and other foreign firms (Felizardo et al., 2003).

In this phase, some components firms started investing abroad, though most of them kept their exporting policy. Indigenous countries realize they need to be close to international OEM's decision centers, supplying good quality products and improving constantly.

In 2002, the auto components sector was thriving. It managed to develop strong process technological competences, it was heavily export oriented and it was investing the development of new competitive areas – namely on research, development and engineering – in order to improve its international competitiveness (INTELI, 2003). However, the auto components industry still needs more investments in research capabilities, new product development and design competences as well as on human resources to be competitive in the global arena.

## 5. Analysis

The origin of the auto industry in Portugal was characterized by a strong importing of foreign components and units and by the lack of foreign direct investment. Although importing activities involve contact with international markets, this first phase is not explained by any internationalization theory analyzed above. Although several innovation activities took place, based on passionate entrepreneurs, a sectoral systemic perspective is totally absent, as the auto industry is totally dependent on foreign companies.

In the 1950s, Portugal lived in economic isolation with a strong industrial protectionism. The imposition of importing quotas attracted foreign investments and imposed the diffusion of the industry at national level. This second phase is marked by the development of the national industry based on an imports substitution policy deployed to promote production activities in Portugal. As a consequence, foreign companies establish their activities in order to expand their market share. Faced with their potential of growth, those firms internalize the Portuguese market based on ownership and location advantages they possess *vis-à-vis* indigenous competitors. It is possible to apply the internalization theory and the eclectic paradigm to explain why foreign firm took a foothold in Portugal. In terms of innovation, dirigiste strategies are followed based on strong policy regulations as Portugal lacks technological infrastructures.

The "Protected Market phase" is marked by strong imports substitution policies. Auto manufacturers were governed by a specific and restricted program of national incorporation. As a consequence, there is a proliferation of assembly units, based on foreign direct investments and on licensing contracts. Auto components firms were not involved on this legal framework. The stream of investments in this phase can be explained by ownership and location advantages of the eclectic paradigm (based on an inward perspective, as indigenous firms were not yet internationalized). The assumptions of Uppsala's evolutionary theory are not explicit in this phase. As in the previous phase, Portugal Is still a very closed country in economic terms. Despite all efforts of national authorities to attract investment and to underpin its indigenous supply base, Portugal still follows a dirigiste perspective. All foreign investments so far attracted only account for a very narrow innovation system perspective.

In the 1980s, the auto industry began its consolidation. By that time, the sectoral policy was marked by exports promotion, with incentives to foreign firms to invest in Portugal, based on quality of human resources, labor comparative advantages and geographical location. Multinational firms operating in Portugal underpinned their exports strategy on their relationships with corporate centers. Accordingly, the network approach could be used to explain, in part, the internationalization of the auto industry. The "Renault Project" had a pervasive influence in the whole value chain, influencing auto components suppliers in their first contacts with a more global perspective of the auto industry. The Portuguese auto industry managed to gain credibility; its industrial fabric was recognized by its production process competence and by its comparative labor advantage. The Renault project underpinned the development of a narrow (sectoral) innovation system based on the relationship throughout the value chain.

Renault's project is very important as it created the condition for a technologically embedded innovation networked system as it worked as a gravitational center for the technological development of the components center. Despite all structural efforts, industrial authorities still follow a dirigiste perspective.

This phase was followed by the "Golden Period of Foreign Direct Investment" – based on heavy direct foreign investments – that paved the way for an unprecedented economic growth in the auto industry as well as in the components industry. The underlying theories explaining the international behavior of foreign firms are the internalization theory, the eclectic paradigm and the network approach.

Finally, the sixth phase was marked by the reopening of the market followed by a complete liberalization. The industrial policy was based on attracting foreign direct investment and in the development of technological and innovation competences. In this period is signed the joint-venture between Ford and Volkswagen – Autoeuropa – and other joint-ventures between foreign and national firms. Autoeuropa's project allowed indigenous suppliers to have broader, international horizons and to network with new players in new international markets. The eclectic paradigm and the internalization theories are the most interesting theories to explain the foreign firms' internationalization. On the other hand, the network theory gains support, due to the internationalization of the auto industry base of suppliers.

Autoeuropa's project complemented Renault's in such a way that – and taking into account all foreign investments – the components industry managed to be involved on a regional networked innovation system, based on Autoeuropas's supply chain. Portuguese authorities have heavily invested on institutions that are able to supply strong technological demands at international level.

Table 8 synthesizes the findings throughout the different phases.

| | International performance | Main theories of internationalization | Home country specific factors for foreign firms | Industry-specific factors influencing operations and market behavior | Impact on indigenous firms | Innovation System perspective |
|---|---|---|---|---|---|---|
| **Phase 1 The special EDFOR case (1937-1952)** | • Essentially domestic<br>• Inward oriented | • Imports substitution<br>• Internalization theory and the eclectic paradigm with an inward focus | • Size of domestic market<br>• Low development of domestic market<br>• Role of government | • Lack of technological development<br>• The beginning of product/industry cycle<br>• Weak supplier base | • "Follow the herd" reaction<br>• No network externalities<br>• Opportunistic strategies | • No innovation system<br>• Portugal as a grassroots country |
| **Phase 2 Automotive Assembly Lines (1950s to 1960s)** | • Essentially domestic<br>• Inward oriented | • Imports substitution<br>• Internalization theory and the eclectic paradigm with an inward focus | • Role of government<br>• "Follow the herd" reaction<br>• Foothold on domestic markets<br>• Opportunistic strategies | • Lack of technological development<br>• Product/industry cycle in its infancy<br>• Search for first mover advantage<br>• Foreign firms as an opportunity<br>• Weak supplier base<br>• Imports seen as a threat | • "Follow the herd" reaction<br>• No network externalities<br>• Learning phase | • No innovation system<br>• Portugal as dirigiste region |
| **Phase 3 Protected Market (1961-1974/76)** | • Essentially domestic<br>• Inward oriented | • Imports substitution<br>• Internalization theory and the eclectic paradigm with an inward focus | • Booming of assemblers activities<br>• Adapting product to local markets<br>• No involvement in upstream activities in the supply chain | • Increasing product technological development<br>• Product/industry cycle in its infancy<br>• Inefficient indigenous suppliers<br>• Foreign firms as an opportunity | • Search for first mover advantage<br>• Some network externalities<br>• Size of domestic market<br>• Learning phase | • No innovation system<br>• Portugal as dirigiste region<br>• Narrow system of innovation |
| **Phase 4 The Renault Project (1977-1986/88)** | • Winds of change<br>• Inward and outward oriented<br>• The beginning of internationalization | • Exports promotion<br>• Internalization theory and the eclectic paradigm with an inward focus<br>• Network approach | • Restructuring perspectives for assemblers<br>• Adapting product to local and international markets<br>• Increasing relationships suppliers in upstream activities in the supply chain<br>• Auto components industry increases in importance | • Product/industry cycle in its growing stage<br>• Growing market<br>• Portugal as exporting platform<br>• Strong comparative labor advantages<br>• Good product quality | • Strong production process technological development<br>• Qualification of human resources<br>• Good product quality<br>• Strong technology and organizational learning<br>• Good network externalities | • Territorially embedded regional innovation networks<br>• Portugal as dirigiste region |

Table 8. Continued

| | International performance | Main theories of internationalization | Home country specific factors for foreign firms | Industry-specific factors influencing operations and market behavior | Impact on indigenous firms | Innovation System perspective |
|---|---|---|---|---|---|---|
| **Phase 5 The Golden Period of Foreign Investment (1987)** | • Inward and outward oriented • Changing period | • Eclectic paradigm with an inward focus • Network approach | • Deep restructuring of auto and components industry • International market perspective • Integrative perspective of auto and components industry | • Economic opening phase • Growing (European) market • Portugal as exporting platform | • Outward perspective • Good product quality • Strong technology and absorptive capacity • Good network externalities | - |
| **Phase 6 Autoeuropa Project (1989-2011)** | • Outward oriented • Outward seeking internationalizati on | • Network approach | • Auto and components industry immersed on a global perspective • Product focus on international markets • Holistic perspective of the value chain | • European industry perspective • Growing market • Portugal as exporting platform • Strong comparative labor advantages • Good product quality | • Strong product and process technological development • Qualification of human resources • Good product quality • Strong absorptive capacity • Strong network externalities | • Portugal in search of a networked region • Regional networked innovation system |

Table 8. Synthesis of internationalization and systemic perspective

## 6. Conclusions and limitations

As mentioned in the introduction, the main objective of this chapter was to analyze how the main theories of internationalization are related to the milestones of the automotive innovation system as well as, the factors that led to the internationalization of the auto industry.

There are eight phases that influenced the internationalization of the Portuguese auto industry. It is clear that the inward perspective of the automobile industry paved the way for the outward perspective of the components industry. From a closed economy with weak endogenous (technological, managerial, strategic and operational) capabilities, Portugal managed to evolve throughout time to a competitive position in the international auto industry arena.

Departing from an absent innovation systemic perspective, Portugal faced the difficulty of creating endogenous technological conditions in ill endowed industries. As described in the previous sections, foreign direct investments underpinned the transfer of technological competences to the Portuguese auto industry as well as the components industry. It was not an easy task as it involved the development of strong absorptive capabilities of indigenous firms.

The path towards more international perspective also needs a strong involvement in the creation of a dynamic innovation system that underpins the efficiency of all the players of the system.

Although Portugal is still strongly dependent on Autoeuropa's factory, and international comparisons could dictate Portugal is still lagging behind their European competitors – with stronger resources and capabilities in the auto industry –, it is possible to conclude that the internationalization of the economy/industry plays a role in the development of the innovation system. This can be explained by the networking activities most players have with international, more demanding players, which fosters the development of competitive advantages based on strong absorptive capabilities.

The main limitation of this research is that it was based on a single country and followed a historical perspective. Future research should address, thus, the relationship between the country's degree of openness and the degree of its innovatory capacity.

## 7. References

Andersen, O. (1993). On the Internationalization of the Firm: a Critical Analysis. *Journal of International Business Studies*, Vol. 24 No. 2, pp. 209-31.

Anderson, E. & Coughlan, A. (1987). International Market Entry and Expansion via Independent or Integrated Channels of Distribution. *Journal of Marketing*, Vol. 51, January, pp. 71-82.

Anderson, E. & Gatignon, H. (1986). Modes of Foreign Entry: a Transaction Costs Analysis and Propositions. *Journal of International Business Studies*, Vol. 17 No. 3, pp. 1-25.

Arocena, R. & Sutz, J. (2000). Looking at National Systems of Innovation from the South, *Industry and Innovation*, Vol. 7, pp. 55-75.

Asheim, B. & Isaksen, A. (2002). Regional Innovation Systems: The Integration of Local Sticky and Global Ubiquitous Knowledge, *Journal of Technology Transfer*, Vol. 27, No. 1, pp. 77-86.

Benito, G. & Welch, L. (1997). De-internationalisation, *Management International Review*, Vol. 37, No. 2, pp. 7-25.

Breschi, S. & Malerba, F. (1997). Sectoral Systems of Innovation: Technological Regimes, Schumpeterian Dynamics and Spatial Boundaries, In: Edquist, C. (Ed.) (1997). *Systems of Innovation*, Printer, London.

Buckley, P.J. & Casson, H. (1976). *The Future of the Multinational Enterprise*. MacMillan, London.

Buckley, P.J. & Ghauri, P.N. (1993). Introduction and Overview, In: Buckley, P.J. & Ghauri, P.N. (Eds.). *The Internationalization of the Firm: A Reader*, Academic Press, London, pp. ix-xxi.

Buckley, P.J.; & Casson, M. (1993). A Theory of International Operation, In: Buckley, P.J. & Ghauri, P.N. (Eds.). *The Internationalization of the Firm: A Reader*, Academic Press, London, pp. 45-50.

Calof, J. & Beamish, P. (1995). Adapting to Foreign Markets: Explaining Internalisation, *International Business Review*, Vol. 4, No. 2, pp. 115-31.

Carlsson, B. & Stankiewicz, R. (1995). On the Nature, Function and Composition of Technological Systems, In: Carlsson, B. (1995). *Technological Systems and Economic Performance – The Case of Factory Automation*, Kluwer, London.

Carlsson, B. (1995). *Technological Systems and Economic Performance – The Case of Factory Automation*, Kluwer, London.

Carlsson, B.; Jacobsson, S.; Holmén, M. & Rickne, A. (2002). Innovation Systems: Analytical and Methodological Issues, *Research Policy*, Vol. 21, pp. 233-245.

Casson, M. (1979). *Alternatives to the Multinational Enterprise*. Macmillan, London.

Caves, R.E. (1982). *Multinational Enterprise and Economic Analysis*. Cambridge University Press, Cambridge, MA.

Chetty, S. & Hunt, C.C. (2003). Paths to Internationalisation Among Small-to Medium-sized Firms, *European Journal of Marketing*, Vol. 37, No. 5/6, pp. 796-820.

Chetty, S. (1999). Dimensions of Internationalisation of Manufacturing Firms in the Apparel Industry, *European Journal of Marketing*, Vol. 33, No. 1/2, pp. 121-42.

Cooke, P. & Memedovic, O. (2003). *Strategies for Regional Innovation Systems: Learning Transfer and Applications*, Policy Papers, Vienna: UNIDO.

Cooke, P. & Morgan, K. (1998). *The Associational Economy: Firms, Regions, and Innovation*, Oxford University Press, Oxford.

Cooke, P.; Gómez Uranga, M. & Etxebarria, G. (1997). Regional Systems of Innovation: Institutional and Organisational Dimensions, *Research Policy*, Vol. 26, pp. 474-491.

Denekamp, J.G. (1995). Intangible Assets, Internalization and Foreign Direct Investment in Manufacturing. *Journal of International Business Studies*, Vol. 26 No. 3, pp. 493-504.

Dunning, J.H. (1980). Toward an Eclectic Theory of International Production: Some Empirical Tests. *Journal of International Business Studies*, Vol. 11, No. 1, pp. 9-31.

Dunning, J.H. (1988). The Eclectic Paradigm of International Production: a Restatement and Some Possible Extension. *Journal of International Business Studies*, Vol. 19 No. 1, pp. 1-31.

Edquist, C. (1997). *Systems of Innovation. Technologies, Institutions and Organizations.* Printer, London.

Felizardo, J.R.; Selada, C. & Videira, A. (2003). *Nunca É Cedo para Mudar Portugal.* Edição INTELI, Lisbon.

Féria, L.P. (1999). *A História do Sector Automóvel em Portugal (1895 - 1995).* Gabinete de Estudos e Prospectiva Económica do Ministério da Economia, Lisboa.

Freeman, C. (1987). *Technology Policy and Economic Performance: Lessons from Japan.* Pinter, London.

Gilroy, B.M. (1993). *Networking in Multinational Enterprises: The Importance of Strategic Alliances,* University of South Carolina Press, Columbia, SC.

Gregersen, B. & Johnson, B. (2002). Learning Economies, Innovation Systems and European Integration, *Regional Studies,* Vol. 31, pp. 479-490.

Hansson G., Sundell H. & Ohman M. (2004). The New Modified Uppsala Model – Based on an Anomalistic Case Study at Maleberg Water AB. Bachelor dissertation, Department of Business, Kristianstad University, Sweden.

Heidenreich, M. (2004). The Dilemmas of Regional Innovation Systems, In: Cooke, P.; Heidenreich, M. & Braczyk, H. J. (Eds.) *Regional Innovation Systems.* Routledge, London.

Hennart, J.F. (1982). *A Theory of Multinational Enterprise.* University of Michigan Press, Ann Arbor, MI.

Hennart, J.F. (1989). Can the New Forms of Investment Substitute for the Old Forms? A transaction cost perspective. *Journal of International Business Studies,* Vol. 20, No. 2, pp. 211-233.

INTELI (2003). *A Indústria Automóvel - Realidades e Perspectivas.* INTELI - Inteligência em Inovação e CEIIA - Centro para a Excelência e Inovação na Indústria Automóvel.

Johanson, J. & Mattsson, L.G. (1988). Internationalization in Industrial Systems - a Network Approach, In: Hood, N. & Vahlne, J.-E. (Eds). *Strategies in Global Competition,* Croom Helm, London, pp. 287-314.

Johanson, J. & Mattsson, L.G. (1992). Network Positions and Strategic Action - An Analytical Framework, In: Axelsson, B. & Easton, G. (Eds.) *Industrial Networks: A New View of Reality.* Routledge, London, pp. 205-217.

Johanson, J. & Vahlne, J-E. (1977). The Internationalization Process of the Firm – a Model of Knowledge Development and Increasing Foreign Market Commitments. *Journal of International Business Studies,* Vol. 8 No. 1, pp. 23-32.

Johanson, J. & Vahlne, J-E. (1990). The Mechanism of Internationalization. *International Marketing Review,* Vol. 7 No. 4, pp. 11-24.

Johanson, J. & Vahlne, J-E. (1992). Management of Foreign Market Entry. *Scandinavian International Business Review,* Vol. 1 No. 3, pp. 9-27.

Johanson, J. & Wiederscheim-Paul, F. (1975). The Internationalization of the Firm – Four Swedish Cases, *Journal of Management Studies,* Vol. 12, No. 3, pp. 305-22.

John, G. & Weitz, B. (1988). Forward Integration into Distribution: Empirical Test of Transaction Cost Analysis. *Journal of Law Economics and Organization,* Vol. 4, Fall, pp. 121-39.

Kamien, M. & Schwartz, N. (1982). *Market Structure and Innovation.* Cambridge University Press, Cambridge.

Klein, S., Frazier, G.L. & Roth, V.J. (1990). A Transaction Cost Analysis Model of Channel Integration in International Markets. *Journal of Marketing Research*, Vol. XXVII, May, pp. 196-208.

Lundvalll, B.-Å. (1992). *National Systems of Innovation: Towards a Theory of Innovation and Interactive Learning*. Pinter, London.

Malerba, F. (2002). Sectoral Systems of Innovation and Production, *Research Policy*, Vol. 31, pp. 247-264.

Malerba, F. (2004). Public Policy and the Development and Growth of Sectoral Systems of Innovation. *Beijing Globelics Conference*.

Moreira, A. C. (2005). *Acumulação Tecnológica nas PME no Relacionamento com Multinacionais*, Publismai, Maia.

Moreira, A. C. (2009). The Evolution of Internationalization: Towards a New Theory?, *Economia Global e Gestão*, Vol. 14, No. 1, pp. 41-61.

Moreira, A. C.; Carneiro, L. F. & Selada, C. (2008). Defining the Regional Innovation Strategy for the Year 2015: the Case of the ITCE Clusters in the North of Portugal, *International Journal of Innovation and Regional Development*, Vol. 1, No. 1, pp. 66-89.

Moreira, A. C.; Carneiro, L. F. & Tavares, M. (2007). Critical Technologies for the North of Portugal in 2015: the Case of ITCE Sectors - Information Technologies, Communication and Electronics, *International Journal of Foresight and Innovation Policy*, Vol. 3, No. 2, pp. 187-206.

Morgan R. & Katsikeas C. (1997). Theories of International Trade, Foreign Direct Investment and Firm Internationalization: a Critique. *Management Decision*, Vol. 35, No. 1, pp. 68–78.

Nelson, R. & Rosenberg, N. (1993). Technical Innovations and National Systems, In: Nelson, R. (Ed.) *National Innovation Systems – A Comparative Study*. Oxford University Press, Oxford.

Nelson, R. (1993). *National Innovation Systems – A Comparative Study*, Oxford University Press, Oxford.

Porter, M. (1980). *Competitive Strategy*. The Free Press, New York.

Porter, M. (1990). *The Competitive Advantage of Nations*. The Free Press, New York.

Pugel, T.A. (1978). *International Market Linkages and U.S. Manufacturing: Prices, Profits, and Patterns*. Ballinger, Cambridge, MA.

Rondé, P. & Hussler, C. (2005). Innovation in Regions: What does Really Matter?, *Research Policy*, Vol. 34, pp. 1150-1172.

Rundh, B. (2003). Rethinking the International Marketing Strategy: New Dimensions in a Competitive Market. *Marketing Intelligence & Planning*, Vol. 21, No. 4, pp. 249-257.

Ruzzier, M.; Hisrich, R.D. & Antoncic, B. (2006). SME Internationalization Research: Past, Present and Future. *Journal of Small Business and Enterprise Development*, Vol.13, No. 4, pp. 476-497.

Selada, C. and Felizardo, J. R. (2002). Da Produção à Concepção - Meio Século de História Automóvel em Portugal, In Heitor, M., Brito, J. and Rollo, M. (Eds.), *Engenho e Obra*. Dom Quixote, Lisbon.

Teece, D.J. (1986). Transactions Cost Economics and the Multinational Enterprise. *Journal of Economic Behavior and Organization*, Vol. 7, pp. 21-45.

Volswagen Autoeuropa (2008) Factos e Números, In: *Autoeuropa*, 2010, Avalilable from: <www.autoeuropa.pt/articles/factos-números>

Welch, L.S. & Luostarinen, R. (1988). Internationalisation: Evolution of Concept, *Journal of General Management*, Vol. 14, pp.34-55.

Williamson, O. (1975). *Markets and Hierarchies: Analysis and Antitrust Implications*. Free Press, New York, NY.

Wolf, B.M. (1977). Industrial Diversification and Internationalization: Some Empirical Evidence. *Journal of Industrial Economics*, Vol. 26, December, pp. 177-191.

# 8

# Trade, SBTC and Skill Premia – A Cross-Country and Cross-Gender Analysis

Oscar Afonso*, Alexandre Almeida and Cristina Santos
*CEF.UP, Faculty of Economics, University of Porto, Porto*
*Portugal*

## 1. Introduction

A recent study by Autor et al. (2008) has pointed out that the skill premia has risen in the US since the 60s. Several other authors have highlighted these trends across different OECD countries (e.g., Katz and Murphy, 1992; Machin, 1996; Goldin and Katz, 1999; Chay and Lee, 2000, Conte and Vivarelli, 2007). A common consensus points to the on-going growth of the demand for high-skilled workers, of which the Skill-Biased Technical Change (SBTC) and International Trade (IT) are the often cited sources.

According to the SBTC explanation, technology has evolved following a biased path towards more skilled workers. The bias makes technology complementary by nature to skilled workers and substitute of unskilled, hence expanding the relative productivity and demand for more educated workers (e.g, Bound and Johnson, 1992; Berman et al., 1994; Autor et al., 1998; Acemoglu, 1998; Berman et al., 1998; Galiani and Sanguinetti, 2003; Conte and Vivarelli, 2007). The IT explanation is based on the Stolper-Samuelson theorem's insights according to which IT would lead to the specialization of developed countries in more skill-intensive activities, thus raising the relative demand for skilled workers and the skill premia (e.g., Leamer, 1994; Sachs and Shatz, 1994; Wood, 1994; Feenstra and Hanson, 1996; Borjas et al., 1997; Leamer, 1998; Galiani and Sanguinetti, 2003; Gonzaga et al., 2006).

Even though the debate has been fierce and pending towards SBTC, literature on wage inequality has somewhat ignored how SBTC and IT impact on the skill premia across gender. In fact, even though several studies have suggested that SBTC fails to explain many aspects of the wage-structure changes, namely, the evolution of the skill premia across gender (e.g., Blau and Kahn,1997; Card and DiNardo, 2002; Acemoglu, 2003; Autor et al., 2008; Bryan and Martinez, 2008), empirical analysis are rare. The impact of IT on the gender-related wage inequality also remains unclear and has only been approached by a few authors (e.g., Seguino, 1997). Indeed, surveying empirical literature on the skill premia we observe that gender-related skill premia differential has been subject to a minor attention with particular relevance in determining the universal character of SBTC as an explaining factor.

The goal of this chapter is to contribute to this issue empirically testing for 25 OECD countries how both SBTC and IT have affected the observed skill premia across gender. We

* Corresponding Author

use two direct measures of the skill premia differential between male and female workers, namely, wage ratios per education level. Our estimation results indicate that SBTC conveys a dominant effect over the wage premium on the sample as a whole and for technological leaders, suggesting that in countries where technological intensive production activities are a small part, absorptive capacity may be limited and SBTC is actually not pervasive. IT has a smaller effect on the sample as a whole and for technological leaders; it is however the dominant for followers and always significant. In what concerns the gender-related inequality, we conclude that SBTC has also a strong and symmetric impact on the wage differential (positive on the club of leaders and negative on the club of followers). IT is again relatively less important in the wage gender-differential evolution.

We organized this chapter as follows: in section 2 we conduct our literature review, followed by the model's specification and methodological issues in section 3. Section 4 presents an analysis the estimated results. Section 5 concludes with some concluding remarks.

## 2. Wage premium: Reviewing the empirical literature

In this section we provide an overview on the empirical literature on the skill premia, focusing on empirical literature that has addressed the issue of gender inequality.

### Empirical literature review on the skill premia

Empirical literature on the skill premia has usually debated which approach, SBTC or IT, was more appropriate in explaining the widening of the wage gap between skilled and unskilled workers. Our survey shows that few studies have considered both explanations together. Indeed, the majority of the empirical studies look only at one side of the debate, SBTC or IT.

In line with the Stolper-Samuelson theorem predictions, Wood (1994) concludes that IT contributes to an increase in the skill premia in the developed world and a decrease in the developing world. In a subsequent study, Sachs and Shatz (1996) concluded in the same direction of Wood (1994), finding a link between the increase in IT flows and the skill premia. Using aggregate data for the US manufacturing between 1972 and 1990, Feenstra and Hanson (1996) conclude that outsourcing, proxied by the imports of intermediate inputs contributed significantly to the relative increase in the demand for skilled labour. Leamer's (2001) results on the evolution of wages of productive and non productive workers in the 70s and the 80s in the US indicated that IT had a significant impact in the decline in the relative demand of unskilled labour and thus in the rise of the wage premium.

In an update study, Feenstra and Hanson (2003) estimate IT to be responsible for 15% to 24% of the wage-premium change and SBTC for 8% to 13%. Also Green et al. (2001) identified a positive shift in the demand for skilled labour and in wage inequality in Brazil. However, Gonzaga et al. (2006) contradict the findings of Green et al. (2001). Recently there is a vibrant revived literature on IT and wage inequality, which reveals the increase in interest to the topic: e.g., Amiti and Konings (2007); Goldberg and Pavcnik (2007); Amity and Davis (2008); Broda and Romalis (2008); Helpman et al. (2008); Krugman (2008); Verhoogen (2008); Burstein and Vogel (2009); Egger and Kreickemeier (2009); Goldberg et al. (2008). In particular, the last two empirical studies provide strong evidence showing that imports of intermediates improve technological progress and thus productivity and wages in developing countries.

The largest set of studies in the 90s has focussed on testing SBTC. One such example is Machin and Van Reenen (1998), who studied SBTC on 7 OECD countries finding evidence of a crucial association of R&D intensity and the share in employment of skilled workers. Based on a sample of 12 OECD countries, Berman et al. (1998) concluded that there was a rise in the share of skilled workers across all countries, reinforcing the argument for SBTC's pervasive nature. The study by Katz and Autor (1999) also suggested that SBTC played a major role in explaining the wage-premium trend. Using a sample of 37 countries of different income levels, Berman and Machin (2000) obtained similar results. Studies like Goldin and Katz (1996), Bartel and Sicherman (1999), Kahn and Lim (1998) and Autor et al. (1998) study the US case and concluded in favour of the positive effect of SBTC on the skill premia.

Recently, Autor et al. (2003) estimated a within industry shift in favour of cognitive tasks linked to the increase in computer usage. This shift indicated that SBTC (proxied by computer usage) accounted for 60% of the wage inequality increase in the 70s and the 80s.

The previous literature has provided evidence supporting SBTC's hypothesis. However, other authors have opposed these results, namely Card and DiNardo (2002). Arguing against the SBTC's holistic explanatory power, Card and DiNardo (2002) assessed the wage premium evolution across the US in the 80s and 90s. They found that simultaneously to the expansion of computer usage, the wage skill differential remained stable which contradicts SBTC. Moreover, the authors highlight that SBTC provides no insights on a set of related issues such as the gender-wage gap. Also a recent study by Berman et al. (2005) on India's manufacturing sector devised for the period of 1983 to 1998 revealed inconclusive results regarding SBTC.

Furthermore, a small amount of studies analyze together how IT and SBTC account for the skill-premia evolution. One such study is Esquivel and Rodríguez-López (2003) who study how IT and SBTC impacted over the skill premia between 1988 and 2000; their results point to IT as the most relevant factor, but gender is nowhere mentioned. Manasse et al. (2004) in a study of the evolution of the skill premia in the Italian metal-mechanical industry suggest that IT and SBTC offset each other with SBTC stimulating an increase in wage inequality and IT a reduction. Again, how IT and SBTC impact on gender-based inequality is neglected and also occurs in Melka and Nayman (2004). In this last study, the authors make a comparative analysis of the skill premia in the US and France, concluding that ICT capital deepening and the R&D stock promoted an increase in the demand of college graduates whereas IT seems to be not statistically relevant.

Using a multi-sector version of the Ricardo-Viner model of IT, Blum's (2008) results suggest that the sector bias rather than the skill bias nature of technological change is more relevant. Analysing the US skill premia trend from 1970 until 1996, Blum (2008) concludes that technology has been biased to more technology-intensive sectors.

The above mentioned studies have focussed on the skill-premia analysis disregarding issues of gender-based inequality apart from a reduced number of exceptions, namely Katz and Autor (1999), Card and DiNardo (2002) and Melka and Nayman (2004). Even among these studies, only Card and DiNardo (2002) analyze gender as more than a side question. Chusseau et al. (2008) extensive literature survey supports this conclusion, highlighting the poor attention devoted to how SBTC and/or IT impacted over gender-based wage inequality. Nevertheless, there is a small amount of studies combining skill premia and gender in the analysis. The next sub-section is devoted to a more close analysis of these.

### Empirical literature on the skill premia and gender

Although Acemoglu (1998) stressed the importance of analysing the skill premia also in terms of gender inequality, not many studies have dealt with this issue, which is pointed as one of the possible flaws to the explaining power of the SBTC hypothesis. Nevertheless, even before Acemoglu (1998), Bound and Johnson (1992) analysed the path of wage differentials in the US, considering education levels and gender. Using data from CPS surveys reporting from 1973 to 1988, the authors' results indicate that the skill premia increased and gender inequality decreased during the 80s.

Presenting similar results, Card and DiNardo's (2002) study focuses on the path of wages and the wage structure in the US. In particular, they analyze how wage inequality and the skill premia has evolved across gender, race and age. Analysing the trends observed in the 80s and the 90s, the authors highlight many inconsistencies or at least shortcomings of SBTC explanation. Card and DiNardo's (2002) empirical evidence point to a closing of the gender gap contradicting SBTC theory insights. According to the authors, the education gradient in computer use is higher for men than women and differences in computer use across gender are narrower for higher levels of education attainment. Finally, Card and DiNardo's (2002) evidence also point to a higher use of computers by male college graduates in relation to female ones. Hence, given SBTC's perspective based on computer use/skill complementarily, wage inequality should have widened in high levels of education and closed for the least educated. The SBTC's "rising skill price" perspective would suggest an expansion of inequality across *all* educational levels. Nevertheless, evidence is clear and the wage gap has overall narrowed by 15 percentual points (pp) between 1980 and 1992.

Arguing that computers replace workers performing routine cognitive tasks and manual tasks and that are complementary to workers performing non-routine tasks, Autor et al. (1998) tried to assess the pervasiveness of SBTC across different groups of analysis, including gender. For both men and women, the authors observe significant shifts of the relative demand for skilled workers, though considerably larger for women. However, Autor et al. (1998) do not analyze the impact on the wage premia, but the bigger shift of demand for skilled women may suggest that the gap should diminish at least for highly educated workers.

In general, there are not many empirical studies assessing how IT impacts on wage inequality (e.g., Anderson, 2005). There are even less that deal with gender-related issues. Among the exceptions, we find Seguino (1997) who concludes that the export-led growth of Korea had a small contribution to the narrowing of the gender wage gap. In the same line, Tzannatos (1999) analysis for 12 developing countries covering the 80s and the 90s go in the same direction of Seguino (1997) but estimating a bigger impact. Rama (2001) concludes that in Vietnam there was a significant reduction of the wage gap across gender in all educational levels during Vietnam's IT liberalization.

Galiani and Sanguinneti (2003) have analyzed Argentina's case. Focusing more clearly on the skill premia, they observe that the evolution across gender is actually similar, not addressing this issue from a pure gender gap perspective but providing results that indicate that the skill premia gender gap has increased with Argentina's liberalization.

Oostendorp's (2004) findings suggest that IT and Foreign Direct Investment (FDI) net inflows reduced the gender gap among unskilled workers of developing countries.

In a recent study, Autor et al. (2008) focus on four inequality concepts: changes in overall wage inequality, changes in inequality in the upper and lower halves of the wage distribution, between-group wage differentials and within-group wage inequality. There results point to a narrowing of the gender-wage gaps since 1980.

Bryan and Martinez (2008) show that in the US the trends in male and female income inequality have been similar over the past few decades with interesting aspects, highlighting that the level of inequality seems to be lower among women than men, though increasing in both cases. Across gender, they conclude that inequality has been decreasing, with women's wages catching up with those of men, confirming Autor et al. (2008) results.

To sum up, there is a small number of studies combining the skill premia with gender. Most of them dealt do not focus in explaining asymmetries across gender, nor deal with it as a primary research issue. The surveys of Chusseau et al. (2008) on the gender-gap literature and of Brown and Campbell (2002) on the skill-premia literature highlight the scarce amount of analysis combining both approaches and which Acemoglu (1998) stated, may be an important contribution to the skill premia debate. Hence, we will focus our contribution on this matter, trying to assess how the skill premia evolved across gender, analyzing how IT and SBTC impacted on gender-based inequality.

## 3. Modelling the case for gender wage premium

In the previous section, we showed that empirical studies on the IT versus SBTC debate have neglected gender inequality. Our goal is to test if IT and/or SBTC explain a fall in gender asymmetries per level of education. On follower (developing) countries, IT may result in higher demand for less educated labour and this expansion lead to more equality. Moreover, since women are now the majority of college graduates, SBTC may have contributed to women fulfilling skilled-labour positions, thus reducing wage gaps towards men. On developed countries, wage-premium path across gender may also be explained SBTC, but IT, at least in Stolper-Samuelson theorem's sense, is probably an irrelevant explaining variable.

Based on a sample of 25 OECD countries and for a 10 year period (1997-2006), we estimate the effects of both IT and SBTC on wage-gender inequality. To test our hypothesis, we estimate the following model specification not only for the sample as a whole, but also for a sub-samples of countries defined for an R&D threshold thus trying to stress the mentioned potential differences of kings for different kingdoms:

$$WP_{it} = \theta + \delta_1 SBTC_{it} + \delta_2 IT + \varphi X + v_{it} ,$$

where $i$ and $t$ stands for, respectively the country and year indexes. Moreover, according to our goal and model specification, our sample comprises proxies for the following variables: wage premium across gender, $WP$, skill-biased technical change, $SBTC$, International Trade, $IT$, and income level, $GDP$. To have statistical data coherence we used only OECD databases.

$WP$ stands for wage premium and is our dependent variable. Since we are interested in capturing gender asymmetries and the effects of IT and SBTC, using the data retrieved from

OECD's 2008 Education at Glance Report on the average wages earned by workers with superior, upper secondary and lower secondary education attainments and on wage inequality between men and women, we build skill-premium measures comparing earnings of colleges graduates with the ones of lower secondary graduates for male, $WPM_{s/l}$, female, $WPF_{s/l}$, and we will also use as a third dependent variable wage differential between man and woman per education levels superior, $WPMF_s$, and lower secondary, $WPMF_l$. This set of dependent variables present an advantage over the proxies mostly used in the literature that, usually due to data unavailability, use indirect measures of the skill premia to assess these relationships.

To evaluate the SBTC, several indicators have been used in the literature. For instance, Bartel and Sicherma (1999) used the proportion of scientists and engineers, Autor et al. (1998) used computer usage and Machin and Van Reenen (1998) use the share of R&D expenditures on GDP. We will use this latter option for a set of reasons. It is available at OECD database for the entire period of our analysis and for all the 25 countries, it measures technology, being intimately associated to innovation performance and finally, it is highly correlated to computer usage thus capturing the majority of effects from ICT spread.

To proxy IT openness we follow Thoenig and Verdier (2002). Having retrieved data on imports, exports and GDP, we computed the degree of openness for each country and across time. The higher exposure to IT, should lead to an overall increase in the wage premium for OECD countries since they are relatively high-income countries but may have differentiated impact among them. If there asymmetries are relevant for this set of countries, then we would expect that IT would have a lower relative impact on the college premium on the higher income end of the sample whereas for the other pole, IT should aid reducing inequality.

To assess the existence of asymmetries across different stages of development, we estimate our model for a set of sub-samples resulting from the decomposition of the sample based on the stages of development. In particular, we decomposed the sample according to Castelacci and Archibugi (2008) technological clubs of convergence. They use an algorithm to cluster countries according to two composite factors: technological infra-structures and human capital and codified knowledge creation and diffusion. Their results identify clubs of convergence based on the countries technological capabilities and structural similarities, which, in our view, are particularly adequate to assess the potential asymmetries on the effects of IT and SBTC across countries with different potential technology absorption potential.

We also added a vector $X$, namely, the logarithm of GDP as a control variable.

Table 1 provides a brief statistical summary for the set of used dependent variables.

In particular, Table 1 shows that the wage premium on females is superior in relation to men's when considering wage inequality between college graduates and lower secondary per gender. When comparing the gender wage differential on college worker and on lower secondary workers, it is also observable that the level of gender discrimination is higher on the lower education levels. Some of these issues are further explored on the next section.

Similarly, Tables 2, 3 and 4 summarize statistic analysis for each explaining variable. We just highlight the inclusion of a measure of the annual variability of each explain variable, which will weigh each estimated coefficient thus giving us a more accurate estimated impact.

| Variable | Max | Min | Average | Std Deviation |
|---|---|---|---|---|
| $WPM_{u/l}$ | 3.453 | 0.934 | 1.957 | 0.478 |
| $WPF_{u/l}$ | 3.600 | 1.367 | 2.074 | 0.489 |
| $WPMF_u$ | 2.732 | 0.884 | 1.51 | 0.244 |
| $WPMF_l$ | 2.364 | 1.051 | 1.64 | 0.281 |
| $WP_{u/l}$ | 5.18 | 0.95 | 1.97 | 0.51 |

*Statistical Source: OECD Science and Technology Indicators.*

Table 1. Statistical summary of the variables used in the model's estimation.

| Variable: SBTC | Max | Min | Average | Std Deviation | $\overline{\Delta SBTC}$ |
|---|---|---|---|---|---|
| Full Sample | 4.77 | 0.49 | 1.90 | 0.96 | 0.03 pp |
| Leaders | 4.77 | 0.99 | 2.35 | 0.81 | 0.03 pp |
| Followers | 1.47 | 0.29 | 0.92 | 0.25 | 0.02 pp |

*Statistical Source: OECD Science and Technology Indicators.*

Table 2. Statistical Summary on SBTC - proxied by the annual share of R&D expenditures on GDP.

| Variable: IT | Max | Min | Average | Std Deviation | $\overline{\Delta TRADE}$ |
|---|---|---|---|---|---|
| Full Sample | 164.10 | 13.88 | 47.72 | 26.54 | 1.47 pp |
| Leaders | 164.10 | 14.84 | 51.83 | 26.98 | 1.24 pp |
| Followers | 100.74 | 13.88 | 39.50 | 23.73 | 1.55 pp |

*Statistical Source: OECD Science and Technology Indicators.*

Table 3. Statistical Summary on International Trade - proxied by the degree of openness.

| Variable: LnGDP | Max | Min | Average | Std Deviation | $\overline{\Delta LnGDP}$ |
|---|---|---|---|---|---|
| Full Sample | 16.39 | 11.15 | 12.96 | 1.14 | 0.05 |
| Leaders | 16.39 | 11.15 | 13.07 | 1.23 | 0.05 |
| Followers | 14.36 | 11.31 | 12.73 | 0.89 | 0.05 |

*Statistical Source: OECD.*

Table 4. Statistical Summary on LnGDP.

To control for differences in structural characteristics, we use Castellaci and Archibuggi's (2008) clubs of convergence which splits our sample into leaders and followers in technological terms. Table 5 indicates the composition of each different clubs.

| Full Sample (N=25) | All countries |
|---|---|
| Technology Convergence Club: Leaders (N=17) | Australia, Austria, Belgium, Canada, Denmark, Finland, France, Germany, Korea, Netherlands, New Zealand, Norway, Sweden, Switzerland, United Kingdom, United States, Israel. |
| Technology Convergence Club: Followers (N=8) | Czech Republic, Hungary, Ireland, Italy, Poland, Portugal, Spain, Turkey. |

Table 5. Composition of each sample group.

## 4. Cross country evidence on the explanatory degree of SBTC and International Trade

In this section we present and analyze the estimation results. In particular, we analyze in what way the two main explanations for the increase in wage inequality per education attainment, SBTC and IT, have in fact promoted the skill premia across gender and how they have impacted in gender-based inequality in both high and low level of education attainment. Next we analyze the estimates derived per each dependent variable and Table 6 presents our model estimation results assessing the evolution of the skill premia on male and female workers.

| Variables | $WPM_{u/l}$ | | | $WPF_{u/l}$ | | |
|---|---|---|---|---|---|---|
| | all | leaders | followers | all | leaders | followers |
| SBTC | 0.26200*** | 0.30194*** | -0.11271 | 0.29136*** | 0.29516*** | -0.15598 |
| Weighed effect | 0.00786*** | 0.00906*** | -0.00236 | 0.00874*** | 0.00885*** | -0.00312 |
| IT | 0,00283** | 0.00382*** | -0.00989*** | 0.00330*** | 0.00312*** | 0.00411*** |
| Weighed effect | 0.00416** | 0.00474*** | -0.01533*** | 0.00485*** | 0.00387*** | 0.00637*** |
| LnGDP | -0.00911 | 0.05793 | -0.35607*** | -0.14123*** | 0.03627 | -0.25405*** |
| Constant | --- | --- | 7.28898*** | --- | --- | 5.63096*** |
| NT | 250 | 170 | 80 | 250 | 170 | 80 |
| Adjusted R² | 0.934 | 0.879 | 0.328 | 0.95633 | 0.934 | 0.960 |
| Method | FEM | FEM | Pool OLS | FEM | FEM | REM |

Notes: ***Significant at 1%, ** significant at 5%, * significant at 10%; Effects: Group only (G) or Group and Time (G&T). Following Wooldridge (2002), we use the global significance F-test, the Lagrange-Multiplier (LM) test and the Hausman test to choose which model (pool OLS, FEM or REM) is more suitable for our estimation in each case.

Table 6. Panel data estimation results of wage premium on male and female individuals.

When using the full sample and the club of convergence of more technology advanced countries, both the marginal and weighted effect of SBTC is dominant and positive, hence promoting an increase in the wage premium for both men and women. Nevertheless, it is statistically not significant for follower countries. For the full sample, SBTC has an estimated marginal effect of 26.2 pp over the wage premium of men and of 29.1 pp over the women's. Since the SBTC proxy scale of annual variability, the computed weighted effect amounts to

0.8 pp and 0.9 pp for men and women, respectively. For technologically advanced countries ("leaders"), the results are slightly higher on both male and female workers.

IT is also estimated to increase the wage premium for the full sample and the leaders group. However, the marginal effect is quite small and the weighted effect is about half the one estimated for SBTC. For the overall sample, the weighted effect is estimated to be 0.4 pp on men and 0.5 pp on women, slightly increasing for men in the leaders group and decreasing for women. However, despite SBTC dominance in the full sample and the leaders' group, for the set of countries classified as technological followers, SBTC is estimated to be not significant both for men and women. Here IT impact, both in marginal and weighted terms is dominant and higher than the registered for the other two samples. IT accounts for a marginal decrease in inequality estimated in about -9.9 pp for men but a positive of 4.1 pp for women. In weighted terms, these values would reach -15.3 pp and 6.4 pp, respectively.

Ln*GDP* is significant on followers and for both genders, accounting for a decrease in inequality estimated, in marginal terms, in -0.36 pp for men and -0.25 pp for women. The evidence supporting the impact of income in decreasing inequality is also present for the full sample, but only for women.

Thus, SBTC is dominant on the sample as a whole and for leaders, suggesting that in countries where technological intensive production activities are a small part, absorptive capacity may be limited and SBTC is actually not pervasive. Though IT has a smaller effect on the first two sets of countries, it is dominant for followers and always significant, in spite of symmetrical. In follower countries, IT contributes to a reduction of the wage premium of males but for an increase in inequality on females. Ln*GDP* is only relevant and inequality reducing on follower countries. In terms of impact, GDP leaps seem to promote a higher inequality which we believe may be explained by the predominance of low-tech industries which, in line with Stolper-Samuelson, suffering an expansion from increased openness and low tech specialization, pushes outwards the demand for less skilled workers and hence contributes to the diminishing of the wage premium, in line with the estimations.

Literature has also questioned how SBTC and IT impacted on gender-based inequality (e.g. Acemoglu, 1998). The skill premia between male and female workers is quite significant (see Table 1), with men earning on average more from 51% to 64% more than women with the same education attainment. In here, we attempt to understand if for workers with the same competences (college or lower secondary), the observed gender related wage inequality increases or decreases with SBTC or IT. Hence, we re-estimated our model using the wage ratio of men to women for college educated workers and also for workers with less than a lower secondary degree. Results are synthesized in Table 7.

Analyzing the results on the gender-based wage inequality among college educated workers, the SBTC is the dominant factor in explaining the evolution observed. It is estimated to have a positive marginal impact of 16.4 pp and a weighted impact of 0.49 pp in leaders. In followers, our results indicate a symmetric effect with SBTC actually contributing to a decrease in the wage inequality between genders. This decrease amounts to -61.1 pp in marginal terms and a variability weighted effect of -1.2 pp. However, SBTC is not significant for the sample as a whole, probably be due to the profound symmetries estimated between leaders and followers. These results suggest that SBTC has a biased impact not only across workers skills, but also across genders and the technological development stage of countries.

| variables | WPMF$_u$ | | | WPMF$_l$ | | |
|---|---|---|---|---|---|---|
| | all | leaders | followers | all | leaders | followers |
| *SBTC* | -0.00565 | 0.16399*** | -0.61125** | 0.14806** | 0.17238*** | -0.11079 |
| *Weighed effect* | -0.00017 | 0.00492*** | -0.01223** | 0.00444** | 0.00517*** | -0.00222 |
| *IT* | -0.0131*** | 0.00125** | -0.00542 | 0.00045 | -0.00005 | 0.01008*** |
| *Weighed effect* | -0.01926*** | 0.00155** | -0.00840 | 0.00066 | 0.00006 | 0.015624*** |
| *LnGDP* | 0.01957** | -0.10323*** | 0.17204 | -0.08942 | -0.15995*** | 0.11894*** |
| *Constant* | 1.38517 | --- | --- | --- | --- | -0.28615 |
| **NT** | 250 | 170 | 80 | 250 | 170 | 80 |
| **Adjusted R²** | 0.74 | 0.77 | 0.3 | 0.82 | 0.88 | 0.37 |
| **Method** | Pool OLS | FEM | FEM | FEM | FEM | Pool OLS |

Notes: see Table 6.

Table 7. Panel Data Estimation on gender based wage differential among college graduates and among lower secondary graduates.

IT is estimated to have a negative impact on the wage gap between male and female workers with college degree for the sample as a whole. In particular, our estimates indicate that IT has a marginal effect of -1.3 pp and a weighted effect of -2.0 pp. IT is also a relevant explaining factor for the set of leaders. For these countries, however, the impact of IT is estimated to widen inequality, with a positive, though small, effect of 0.1 pp and a weighted effect of 1.6 pp. In the club of convergence grouping the less technologically advanced countries, IT is estimated to have a negative impact but not statistically significant.

An interesting result arises from our control variable, LnGDP. LnGDP is not significant for follower countries. However, it is estimated to have a positive impact for the sample as a whole and a negative one in leaders. Not only this result seems to indicate that among the most advanced countries, the most gender egalitarian societies have a relatively lower GDP, but it may also become a dominant factor in a context of strong economic growth.

In sum, SBTC is also a predominant explanation for wage inequality widening across genders within college educated workers, conveying a symmetric effect (positive on leaders and negative on followers). IT is a less important determinant of wage inequality however, for college graduates and the sample as a whole, being not statistically significant for followers whereas GDP may become a dominant factor in a context of strong economic growth, contributing to the diminishing of inequalities between gender.

The second set of estimates on Table 7 redoes the above analysis for a set of workers with an educational attainment equal or below lower secondary degrees. Similarly to the results for the set of workers with college degrees, SBTC arises as the dominant explanation except for followers. For the full sample, SBTC accounts for an estimated marginal effect of 14.8 pp and a weighted effect of 0.44 pp. These estimates are slightly higher when we re-estimated the model for the set of leaders. However, despite SBTC's dominance for these groups of countries, in followers our results indicate no statistical significance in spite of indicating also a negative effect upon gender inequality.

Unlike the estimates for workers with a college degree, now IT is only significant for the set of followers where in fact it contributes to an increase in inequality. Here, an increase in the openness level of the economy in 1 pp would result in an increase in the wage differential between men and women of approximately 1 pp in marginal terms.

The symmetric effect of the GDP level on gender-inequality is again present and following the already noticed path in the first set of estimates of Table 7. In particular, the GDP level has an inequality reducing impact on leaders and a widening impact on followers.

In sum, there seems to be a very relevant association of the pervasiveness of SBTC and a country technological development, having SBTC a dominant impact on these countries. Nevertheless, SBTC's effect is apparently symmetrical, contributing to the widening of the wage gap both within gender and across genders on technologically more advanced countries but to a an inequality reduction on less technologically advanced countries.

In what concerns IT, among college graduate workers, the overall effect is small but inequality reducing, signal that is common to the followers' sub-sample. For leaders, IT widens the gender wage gap. When assessing the impact of IT and SBTC hypothesis on the gender wage discrimination among lower skilled workers, SBTC seems to convey a positive impact both on the sample as a whole and on leaders, thus increasing wage inequality not education based. On followers, despite of being statistically insignificant, the sign is negative, as it was observed for college graduate workers. In terms of the fecundity of IT in explaining wage inequality between men and women, IT is only relevant for followers where it contributes positively to the wage differential between genders.

GDP has an inequality reducing effect on the wage premium and also on the wage differential per gender apart from the case of the full sample in college educated workers and followers for the group of less educated workers.

## 5. Conclusions

Literature has long been debating the reasons for the observed increase in the college wage premium focusing on theoretical arguments centered on two explanations, SBTC and IT. Our literature review highlighted the need to further empirical studies, namely assessing the SBTC and IT explanations impact on wages in a different perspective. Hence, we use SBTC and IT to assess based on a 25 OECD countries sample, possible asymmetries and the actual effects on the reduction or widening of gender inequality.

Using the traditional measure of wage premium and comparing differences between males and females, results show that SBTC is overall dominant, with IT, despite its smaller impact, being always significant and in a sense universal in explaining the skill premia.

SBTC's effect is symmetric across clubs of convergence and asymmetric in impact. In terms of signal, SBTC widens the wage premium and the gender differential for the full sample and leaders, apparently promoting a wage inequality decrease common to genders on followers (though, the latter effect is not statistically significant in all estimates). But the effects are also asymmetric in magnitude with SBTC accounting for a stronger impact on technologically more advanced countries, suggesting that the pervasiveness of technology is equal across distinct technological/economic country structural profile. SBTC is also a major explanation for wage inequality widening across genders within the same levels of education skill. Thus, SBTC has a biased impact not only across workers skills, but also across genders.

IT is overall almost always statistically significant, conveying a higher impact over the set of technologically followers. IT estimated impact is lower than SBTC's however there is an interesting symmetric result that points to IT reducing the skill premia among men and widening it among women.

In sum, SBTC arises as the dominant explanation for the evolution of the wage premium within gender and the wage differential between genders, though the effects depend on the stage of development of a country. IT is estimated to convey a less powerful impact on wages although IT is more sample-universal.

## 6. References

Acemoglu, D. (1998). "Why do new technologies complement skills? Directed technical change and wage inequality." *Quarterly Journal of Economics*, vol. 113(4), pp. 1055-1089.

Acemoglu, D. (2003). "Patterns of skill premia." *Review of Economic Studies*, vol. 70(2), pp. 199-230.

Amiti, M. and Konings, J. (2007). "Trade liberalization, intermediate inputs and productivity: evidence from indonesia." *American Economic Review*, vol. 97(5), pp. 1611-1638.

Amiti, M. and Davis, D. (2008). "Trade, firms and wages: theory and evidence." NBER Working Papers No. 14106.

Anderson, E. (2005). "Openness and inequality in developing countries: a review of theory and recent evidence." *World Development*, vol. 33(7), pp. 1045-1063.

Autor, D., Katz, L. and Kearney, M. (2008). "Trends in U.S. wage inequality: revising the revisionists." *Review of Economics and Statistics*, vol. 90(2), pp. 300-323.

Autor, D., Katz, L. and Krueger, A. (1998). "Computing inequality: have computers changed the labour market?" *Quarterly Journal of Economics*, vol. 113(4), pp. 1169-1213.

Autor, D., Levy, F. and Murnane, R. (2003). "The skill content of recent technological change: an empirical exploration." *Quarterly Journal of Economics*, vol. 118(4), pp. 1279-1333.

Bartel, A. P. and Sicherman, N. (1999). "Technology change and wages: an inter-industry analysis." *Journal of Political Economy*, vol. 107(2), pp. 285-325.

Berman, E. and Machin, S. (2000). "Skill-biased technology transfer around the world." *Oxford Review of Economic Policy*, vol. 16(3), pp. 12-22.

Berman, E., Bound, J. and Griliches, Z. (1994). "Changes in the demand for skilled labour within U.S. manufacturing industries: evidence from the annual survey of manufacturing." *Quarterly Journal of Economics*, vol. 109(2), pp. 367-397.

Berman, E., Bound, J. and Machin, S. (1998). "Implications of skill-biased technological change: international evidence." *Quarterly Journal of Economics*, vol. 113(4), pp. 1245-1279.

Berman, E., Somanathan, R. and Hong, T. (2005). "Is skill-biased technological change here yet? Evidence from Indian manufacturing in the 1990." *World Bank Policy Research Working Paper* N° 3761.

Blau, F. and Kahn, L. (1997). "Swimming upstream: Trends in the gender wage differential in the 1980s." *Journal of Labor Economics*, vol. 15(1), pp. 1-42.

Blum, B. (2008). "Trade, technology, and the rise of the service sector: the effects on US wage inequality." *Journal of International Economics*, vol. 74, pp. 441-458.

Borjas, G., Freeman, R., Katz, L., DiNardo, J. and Abowd, J. (1997). "How much do immigration and trade affect labor market out-comes?" *Brookings Papers on Economic Activity*, pp. 1-90.

Bound, J. and Johnson, G. (1992). "Changes in the structure of wages in the 1980's: an evaluation of alternative explanations." *American Economic Review*, vol. 82(3), pp. 371-392.

Broda, C. and Romalis, J. (2008). "Inequality and prices: does China benefit the poor in America?" Working Paper, University of Chicago.

Brown, C. and Campbell, B. (2002). "The impact of technological change on work and wages." *Industrial Relations*, vol. 41(1), pp. 1-33.

Bryan, K. and Martinez, L. (2008). "On the evolution of income inequality in the United States." *Economic Quarterly*, vol. 94(2), pp. 97-120.

Burstein, A. and Vogel, J. (2009). "Globalization, technology, and the skill premium." working paper, UCLA.

Card, D. and DiNardo, J. (2002). "Skill-biased technological change and rising wage inequality: some problems and puzzles." *Journal of labour Economics*, vol. 20(4), pp. 733-783.

Castellacci, F. and Archibugi, D. (2008). "The technology clubs: The distribution of knowledge across nations." *Research Policy*, vol. 37(10), pp. 1659-1673.

Chay, K. and Lee, D. (2000). "Changes in relative wages in the 1980s: returns to observed and unobserved skills and black-white wage differentials." *Journal of Econometrics*, vol. 99(1), pp. 1-38.

Chusseau, N., Dumont, M. and Hellier, J. (2008). "Explaining rising inequality: skill-biased technical change and North-South trade." *Journal of Economic Surveys*, vol. 22(3), pp. 409-457.

Conte, A. and Vivarelli, M. (2007). "Globalization and employment: imported skill-biased technological change in developing countries." *JENA Economic Research Papers*, n° 9.

Egger, H. and Kreickemeier, U. (2009). "Firm heterogeneity and the labor market effects of trade liberalization." *International Economic Review*, vol. 50, pp. 187-216.

Esquivel, G. and Rodríguez-Lopes, J. (2003). "Technology, trade, and wage inequality in Mexico before and after NAFTA." *Journal of Development Economics*, vol. 72, pp. 543- 565.

Feenstra, R. and Hanson, G. (1996). "Globalization, outsourcing, and wage inequality." *American Economic Review*, vol. 86(2), pp. 240-245.

Feenstra, R. and Hanson, G. (2003). "Global production sharing and rising inequality: a survey of trade and wages.", in *Handbook of International Trade*, vol. 1, edited by E. Kwan Choi and James Harrigan, Malden, Mass.: Blackwell Publishing.

Galiani, S. and Sanguinetti, P. (2003). "The impact of trade liberalization on wage inequality: evidence from Argentina." *Journal of Development Economics*, vol. 72, pp. 497-513.

Goldberg, P. and Pavcnik, N. (2007). "Distributional effects of globalization in developing countries." *Journal of Economic Literature*, vol. 45, pp. 39-82.

Goldberg, P., Khandelwal, A., Pavcnik, N. and Topalova, P. (2008). "Imported intermediate inputs and domestic product growth: evidence from India." NBER Working Paper No. 14416.

Goldin, C. and Katz, L. (1996). "Technology, skill and the wage structure: insights from the past." *American Economic Review*, vol. 86(2), pp. 252-257.

Goldin, C. and Katz, L. (1999). "The returns to skill in the United States across the twentieth century." NBER Working Papers No. 7126.

Gonzaga, G., Filho, N. and Terra, C. (2006). "Trade liberalization and the evolution of skill earnings differentials in Brazil." *Journal of International Economics*, vol. 68(2), pp. 345-367.

Green, F., Dickerson, A. and Arbache, J. (2001). "A picture of wage inequality and the allocation of labor through a period of trade liberalization: the case of Brazil." *World Development*, vol. 29(11), pp. 1923-1939.

Greiner, A., Rubart, J. and Semmler, W. (2004). "Economic growth, skill-biased technical change and wage inequality: a model and estimations for the US and Europe." *Journal of Macroeconomics*, vol. 26(4), pp. 597-621.

Helpman, E., Itskhoki, O. and Redding, S. (2008). "Wages, unemployment and inequality with heterogeneous firms and workers." NBER Working Papers No.14122.

Kahn, J. and Lim, J. (1998). "Skilled labor-augmenting technical progress in U.S. manufacturing." *Quarterly Journal of Economics*, vol. 113(4), pp. 1281-1308.

Katz, L. and Autor, D. (1999). "Changes in the wage structure and earnings inequality", in O. Ashenfelter and D. Card (eds.), *Handbook of Labor Economics*, vol. 3. Amsterdam: Elsevier-North Holland.

Katz, L., Murphy, K. (1992). "Changes in relative wages, 1963–1987: supply and demand factors." *Quarterly Journal of Economics*, vol. 107(1), pp. 35-78.

Krugman, P. (2008). "Trade and wages, reconsidered." *Brookings Papers on Economic Activity*, pp. 103-154.

Leamer, E. (1998). "In search of Stolper-Samuelson linkages between international trade and lower wages", in Collins, S. (Ed.), *Imports, Exports and the American Worker*, Brookings Institution, pp. 141-202.

Leamer, E. (2001). "In search of Stolper-Samuelson effects on U.S. wages." NBER Working Papers No 5427.

Leamer, Edward E. (1994). "Trade, wages and revolving door ideas." NBER Working Papers No 4716.

Machin, S. and Van Reenen, J. (1998). "Technology and changes in skill structure: evidence from seven OECD countries." *Quarterly Journal of Economics*, vol. 113(4), pp. 1215-1244.

Machin, S. (1996). "Wage inequality in the UK." *Oxford Review of Economic Policy*, vol. 12(1), pp. 47-64.

Manasse, P., Stanca, L. and Turrini, A. (2004). "Wage premia and skill upgrading in Italy: why didn't the hound bark?" *Labour Economics*, vol. 11(1), pp. 59-83.

Melka, J. and Nayman, L. (2004). "Labour quality and skill biased technological change in France." *Centre d'Études Prospectives e d'Informations Internationales*.

Oostendorp, R. (2004). "Globalisation and the gender wage gap", *World Bank Working Paper* No 3256.

Rama, M. (2001). "The gender implications of public sector downsizing: The reform program of Vietnam." *World Bank Working Paper* No 2573.

Sachs, J. and Shatz, H. (1996). "US trade with developing countries and wage inequality." *American Economic Review*, vol. 86(2), pp. 234-239.

Seguino, S. (1997). "Gender wage inequality and export-led growth in South Korea." *Journal of Development Studies*, vol. 34(2), pp. 102-132.

Thoenig, M. and Verdier, T. (2003). "A theory of defensive skill biased innovation and globalization." *American Economic Review*, vol. 93(3), pp. 709-728.

Tzannatos, Z. (1999). "Women and labour market changes in the global economy: Growth helps, inequalities hurt and public policy matters." *World Development*, vol. 27(3), pp. 551-570.

Verhoogen, E. (2008). "Trade, quality upgrading and wage inequality in the Mexican manufacturing sector." *Quarterly Journal of Economics*, vol. 123, pp. 489-530.

Wood, A. (1994). *North-South trade, employment and inequality*. Oxford University Press.

Wooldridge, J. (2002). *Econometric Analysis of Cross Section and Panel Data*, Cambridge, MA: MIT Press.

# 9

# The Impact of ICTs on Innovative Sustainable Development in East and Southern Africa

Gabriel Kabanda

*Zimbabwe Open University, Harare*
*Zimbabwe*

## 1. Introduction

The major problem of under-development in Africa characterised by the huge challenge to achieve the Millennium Development Goals (MDGs) is on knowledge empowerment supported by Information and Communication Technologies (ICTs). Information has become a strategic resource, a commodity and foundation of every activity. The emergence and convergence of information and communication technologies (ICTs) has remained at the centre of global socio-economic transformations. If implemented properly and carefully, these technologies could reduce or eliminate the imbalance between rich and poor, and powerful and marginalized.

The productive capacity of a country is determined by the quantity and quality of its factors of production. **Infodensity** is the sum of all ICT stocks, mainly as capital and labour. According to UNCTAD (2006), 1% increase in Infodensity resulted on average in 0.3% increase in per capita GDP. The increase in infodensity over time is illustrated below in Figure 1.

Baliamoune-Lutz (2003) conducted research using data from developing countries and examined the links between ICT diffusion and per capita income, trade and financial indicators, education, and freedom indicators. Internet hosts, Internet users, personal computers and mobile phones represent indicators of ICT. It is important to assess the adoption and diffusion of ICTs in key sectors of the economies of Southern and East African countries, and to collate basic information about the actual and potential applications of ICTs in order to have a clear understanding of the specific policy environments and sustainable capacity requirements. Some researchers argue that the transfer of ICTs to developing countries may not contribute to economic development the same way it did in industrial countries, and that it may be best to localize technology and focus on its use in education (Baliamoune-Lutz, 2003) and sustainable development of economic growth.

Diffusion is the process in which an innovation is communicated through certain channels over time among the members of a social system. It is a special type of communication, in that the messages are concerned with new ideas (Rogers,2003, page 5). Innovations diffuse through a social system explained by the diffusion of innovation theory (Rogers, 2003). Diffusion of innovation is a theory that analyzes, as well as explain, the adaptation of a new innovation. The purpose of this theory to the research is to provide a conceptual paradigm

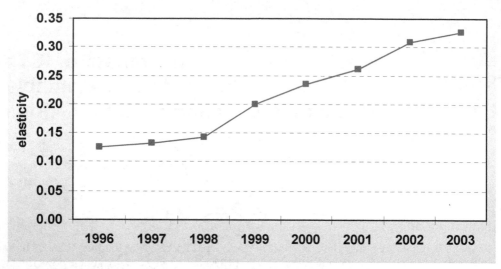

Fig. 1. Infodensity and increase in per capita GDP (UNCTAD, 2006)

for understanding the process of diffusion and social change associated with ICTs. African countries are largely end consumers of technology and fall among the late majority (34%) and laggards (16%) with respect to ICT innovations. This is illustrated by the chart shown below as Figure 2:

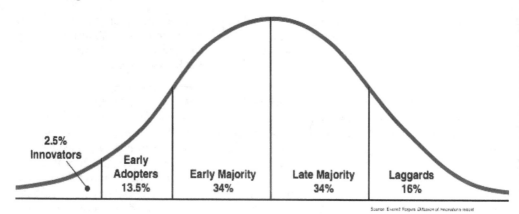

Fig. 2. Rogers' bell curve

Innovations that are perceived by individuals as having greater relative advantage, compatibility, trialability, observability, and less complexity will be adopted more rapidly than other innovations (Rogers, 1995).

The **major developmental problem** being faced by East and Southern African countries, which is the subject of focus for this paper, is multi-faceted and includes the following symptoms:

1.   Many donor-driven initiatives that excluded both policy-formulation frameworks and sustainable capacity building have not brought meaningful development in these African countries.
2.   The Government policies, donor interest and community development needs are totally divergent with respect to priority areas for development.
3.   Africa needs to break the under-development, poverty and illiterate cycles in the long term and exploit the blessed resources available to create wealth. Extensive investment in technology and human capital development as a vehicle to exploit the vast mineral and natural resources has not been given sufficient attention.
4.   Poverty reduction requires a sustainable solution that increases production capacity at individual, institutional, community and national levels. The impact of ICTs on MDGs and generally economic growth needs a detailed assessment.

Some researchers argue that the transfer of ICTs to developing countries may not contribute to economic development the same way it did in industrial countries, and that it may be best to localize technology and focus on its use in education (Baliamoune-Lutz, 2003) and sustainable development of economic growth. The correlation between ICT use and economic growth is interrogated and the issue of the direction of causality is investigated. The **specific objectives of the research** are:

- To assess the impact of ICTs on MDGs
- To ascertain the ICT impact on economic growth, and determine the pattern for diffusion and adoption of ICT innovations in East and Southern Africa
- To recommend a development model or a framework for economic growth for these African countries.

## 2. Literature review

According to the International Telecommunication Union study (ITU, 2010), the relevance and impact of ICTs to the MDGs is tabulated below (http://www.itu.int/ITU-D/ict/publications/wtdr_03/material/ICTs _MDGs.pdf).

Therefore, ICTs impact on all the MDGs in different ways. The fast track to the achievement of MDGs lies greatly in the ability to effectively manage the diffusion and adoption of ICTs for development.

Debates have ensured on how information and communication technologies (ICTs) can help to alleviate poverty in low-income countries (Heeks, 1999). Advances in communication technologies have enabled many countries to improve the lives of their citizens through improved health, education and public service systems, and economies (Kekana, 2002). A knowledge economy requires:

- widespread access to communication networks;
- the existence of an educated labour-force and consumers (human capital); and
- the availability of institutions that promote knowledge creation and dissemination.

A holistic approach to the information economy is required which provides information skills, communication skills and assistance with improving organic-, literate- and intermediate- technology based systems as well as the more obvious ICT-focused areas. A study conducted by Moradi and Kebryaee (2010) explored the impacts of ICT investment on

| MDGs | Impact of ICTs |
|------|----------------|
| MDG 1: Eradicate extreme poverty and hunger | ICTs provide increased access to market information and reduce transaction costs for poor farmers and traders. ICTs create employment and increase wealth. Tele-work allows gainful work from home. ICTs increase skills and productivity resulting in increased incomes. |
| MDG 2 : Achieve universal primary education | ICTs increase supply of trained teachers though ICT-enhanced distance training. Distance learning helps in educational and literacy programmes in rural and remote areas. |
| MDG 3 : Promote gender equality and empower women | ICTs deliver educational and literacy programmes specifically targeted to poor girls and women. Studies show females outnumber males in E-learning programmes. ICTs also empower women to steelwork from home. |
| MDG 4,5,6 : Health (Child mortality, maternal health – reduce by 2/3 and 3/4, HIV AIDS, Malaria, etc. - Halt and reverse ) | ICTs increase access of rural care-givers to specialist support and remote diagnosis. ICTs enhance delivery of basic and in-service training for health workers. ICTs increase monitoring and information-sharing on disease and famine. |
| MDG 7 : Ensure Environmental stability | Remote sensing technologies and communication networks permit more effective monitoring, resource management, and mitigation of environmental risks. Steelwork obviates the need to travel, saves energy and reduces pollution. |
| MDG 8: Global partnership for development | ICTs are extensively used in communication and nurturing of collaborative partnerships. The regional collaboration strategy supported by ICTs covers: Humanware / social issues Software Oriented Technologies Hardware Oriented Technologies E-mail styles and problems Multimedia mail Shared Applications |

Table 1.

economic growth in a cross section of 48 Islamic countries using the data over the period 1995-2005. Panel data analysis was carried out to examine the factors affecting economic growth where the standard Solow growth model was extended to take into account the technological progress, embodied in the form of ICT investment and human capital in order to take the speed of convergence into consideration. The findings showed that the main engines of economic growth are ICT capital, non-ICT capital and human capital in a sample of 48 Islamic countries (Moradi and Kebryaee, 2010), where inflation was noted to have a negative impact on economic growth. ICT investment was found to have a stronger influence on economic growth in the sub-sample of 24 countries that have relatively a higher ICT Opportunity Index. Moreover, non-ICT investment was found to positively affect on economic growth. However, neither openness nor population growth seems to have significant impact on economic growth, although the speed of convergence in both sub-samples was about the same (Moradi and Kebryaee, 2010).

A global set of indicators *(infostate)* showing how the availability of ICTs and access to networks can be a misleading indicator if it neglects people's skills, and if ICT networks and skills combined *(infodensity)* are not matched by a measurement of what individuals, business and countries actually do with such technologies *(info-use)*, is worth further interrogation in this study. This approach offers important perspectives into the central role that e-policies and knowledge have started to play in determining how countries will fare in the global competition to benefit from the information revolution and move away from poverty (Sciadas, 2003). A close correlation exists between Infostates and per capita GDP. Initial study reveals that for every point increase in Infodensity, per capita GDP increases anywhere between $136 and $164 (Sciadas, 2003). The interrelationships between info-state, info-density and info use are illustrated by the diagram shown below as Figure 3 (after Sciadas, 2003, page 10).

# Infodensity and Info-use

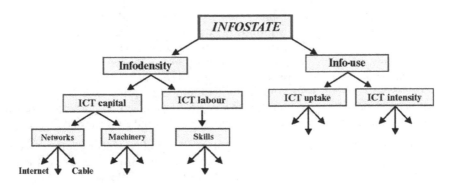

Fig. 3. Infodensity versus Info-use

As a result of the convergence of information, telecommunications, broadcasting and computers, the Information and Communication Technologies (ICTs) sector now embraces a large range of industries and services. The potential of ICTs to transform development is now receiving greater attention worldwide. If ICTs are appropriately deployed, they have the potential to combat rural and urban poverty and foster sustainable development (Samiullah & Rao, 2000). Generally, investment in the development of the ICT infrastructure will result in improved economic efficiency and competitiveness; more efficient and effective education; healthcare and public administration; opportunities to exploit low factor costs in international markets; opportunities to increase social capital; and opportunities to bypass failing domestic institutions.

The African ICT environment and infrastructure faces tremendous challenges, as evidenced by a synopsis conducted for ICT indicators for Africa (Kabanda G., 2008), which shows that:

- Africa has the lowest growth in teledensity of any developing region in the world.
- Has 12% of World population, but 3% of World's main telephone landlines.
- Average level of income is the lowest, but the cost of installing telephone landline is the highest due to the huge costs of civil works involved stretching over very long distances or in areas with a large geographical dispersion.
- Highest profit per telephone landline and long waiting period for telephone lines.
- Internet connectivity is 1.5% of the world-wide connectivity.

Mobile phones may not just help create jobs and new sources of revenue to the state but can also contribute to economic growth by widening markets, creating better information flow, lowering transaction costs, and becoming substitute for costly transportation that is lacking in rural Africa (Kyem, P. A. and LeMaire, P. K. , 2006). The rate of growth of cellphones in Africa has outpaced the growth rate of mainlines. There is a sharp increase in mobile phone growth versus mainlines. The number of mobile subscriptions worldwide is now exceeding five billion, i. e. more people today have access to a cell phone than to a clean toilet.

## 3. Methodology

The research methodology describes ways of obtaining and analyzing data to reach conclusions, thus building up empirical evidence to back up these conclusions. The methodology used was largely qualitative on technology capacity needs assessment that covered 6 countries (South Africa, Kenya, Tanzania, Botswana, Zambia and Zimbabwe), and also quantitative on GDP and Infodensity covering 18 countries in East and Southern Africa. The 18 countries covered by the qualitative study are South Africa, Angola, Bostwana, Burundi, D.R. Congo, Kenya, Lesotho, Madagascar, Malawi, Mauritius, Mozambique, Namibia, Rwanda, Swaziland, Tanzania, Uganda, Zambia and Zimbabwe.

The **quantitative approach** involved the use of surveys and conducting interviews on GDP and Infodensity. The survey method used is good for comparative analysis, got lots of data in a relatively short space of time and was cost-effective. GDP and Infodensity data was collected for 18 African countries to ascertain the link between ICTs diffusion and GDP density per country. Data was collected on **nominal gross domestic product** (GDP) of selected East and Southern African countries, i.e. the market-value of all final goods and services from a nation in a given year. The GDP dollar values presented here were obtained from the IMF (http://www.imf.org/external/pubs/ft/weo/2009/02.weodata/index.aspx) and are calculated at market or government official exchange rates by the International Monetary Fund (IMF) staff. 2009 values and some of 2008 values are estimates. The methodology for collecting data on infodensity is supported by secondary data covering East and Southern African countries. GDP of Africa was 2.5% of the total GDP of the world in 2008, and was estimated to be 2.3% of the world GDP in 2009. In this case,

*GDP Density = GDP per capita * Number of people per square kilometer.*

Data on Infodensity was obtained from the International Telecommunications Union (ITU, 2010).

The **qualitative research** was used to deepen our understanding of the link between diffusion of ICTs and economic growth. It is pleasing to note that the Tripartite Summit signed in October, 2008 provided a political framework for the harmonisation of various policies, initiatives, infrastructure, institutional arrangements and cooperation from the Common Market for East and Southern Africa (COMESA), East African Community (EAC) and the Southern Africa Development Community (SADC) member states.

The technology capacity needs assessment was conducted in institutions and regional bodies in Kenya, Tanzania, South Africa, Botswana, Zambia and Zimbabwe for the period December 2008 to December 2009. Data was collected from government officials, heads of institutions/organizations and experts in various organisations in East and Southern African countries. The capacity needs assessment was conducted in the context of the systems level, the entity level and individual human capital development needs. Capacity needs assessment included both the human capital development and social capital aspects in order to achieve sustainable information and communication technology capacity development. Human capital development is central to capacity needs. A training need exists when there is a gap between what is required of a person to perform their work competently and what they actual know. Interviews were conducted for the organisations and a questionnaire administered in the form of a Capacity Needs Assessment Questionnaire. Secondary data was also compiled and analysed. The following data collection techniques were used in this study:

- Formal meetings and focus group discussions
- Face-to-face oral interviews
- Questionnaires on capacity needs assessment
- Secondary data and records observation

The face-to-face interviews allowed for in-depth knowledge sharing, helped to develop the bigger picture on ICTs for development and was good for networking. Focus group discussions were held with selected regulatory, training and research institutions to pick up grassroots input and in developing ideas, whist sharing latent knowledge spontaneously, on technological capacity needs assessment. Site visits were conducted in selected organizations and institutions. A structured questionnaire, the Capacity Needs Assessment Questionnaire, was administered in the same institutions in order to solicit detailed information in support of the interviews and focus group discussions.

## 4. Analysis of results

### 4.1 GDP for Southern and East Africa countries

The world nominal GDP per capita for the year 2008 was 60,917.477 USD and for the 2009 was 57,228.373. However, the total for Africa was only 1,282.373 (2.11%) in 2008 and 1,184.891 (2.07%) in 2009, respectively. The nominal GDP per capita for the 18 African countries in East and Southern Africa are shown below, with Figure 4 showing the top 8 countries only. Notably South Africa leads all the countries by far, followed by Angola and then Kenya. The lower end in this group has Uganda, Botswana and DRC.

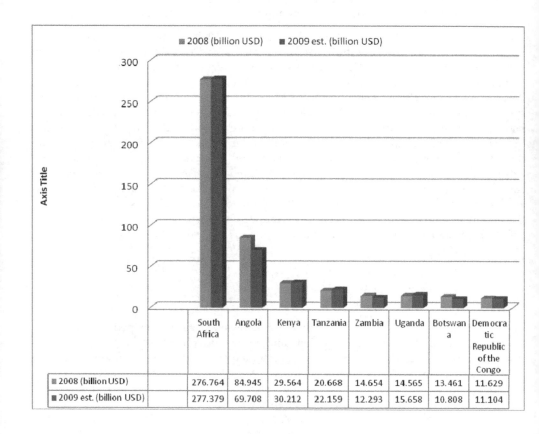

| | South Africa | Angola | Kenya | Tanzania | Zambia | Uganda | Botswana | Democratic Republic of the Congo |
|---|---|---|---|---|---|---|---|---|
| ■ 2008 (billion USD) | 276.764 | 84.945 | 29.564 | 20.668 | 14.654 | 14.565 | 13.461 | 11.629 |
| ■ 2009 est. (billion USD) | 277.379 | 69.708 | 30.212 | 22.159 | 12.293 | 15.658 | 10.808 | 11.104 |

Fig. 4. Nominal GDP per capita for period 2008-2009 for top 8 countries

The diagram below on Figure 5 shows Nominal GDP per capita for period 2008-2009 the bottom 10 countries. This league is led by Mozambique, Madagascar and Mauritius in that order. The lowest 3 in this group are Swaziland, Lesotho and Burundi.

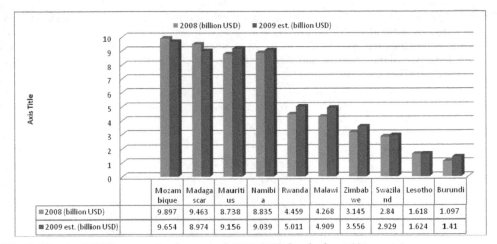

| | Mozam bique | Madaga scar | Mauriti us | Namibi a | Rwanda | Malawi | Zimbab we | Swazila nd | Lesotho | Burundi |
|---|---|---|---|---|---|---|---|---|---|---|
| ■2008 (billion USD) | 9.897 | 9.463 | 8.738 | 8.835 | 4.459 | 4.268 | 3.145 | 2.84 | 1.618 | 1.097 |
| ■2009 est. (billion USD) | 9.654 | 8.974 | 9.156 | 9.039 | 5.011 | 4.909 | 3.556 | 2.929 | 1.624 | 1.41 |

Fig. 5. Nominal GDP per capita for period 2008-2009 for the least 10 countries.

The nominal GDP per capita excluding South Africa is shown in Figure 6 below. Outside South Africa, Angola is a leader followed by Kenya.

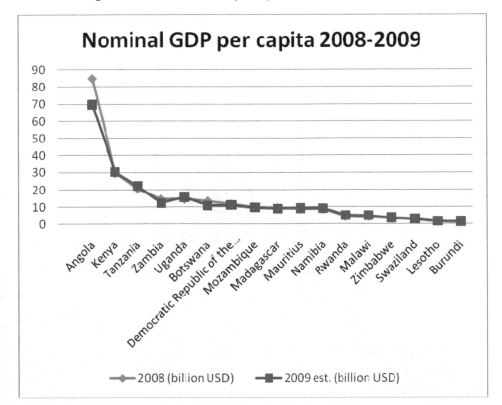

Fig. 6. Nominal GDP per capita for 2008-2009 excluding South Africa.

The GDP per person of the 18 African countries for the period 2008 to 2009, converted to US dollar through estimated IMF exchange rates, are shown below, with Figure 7 below showing the top 8 countries only. Botswana, Mauritius and South Africa are the top 3 in that order whilst Swaziland, Zambia and Kenya are the last 3 among the top 8 countries with respect to the GDP per person.

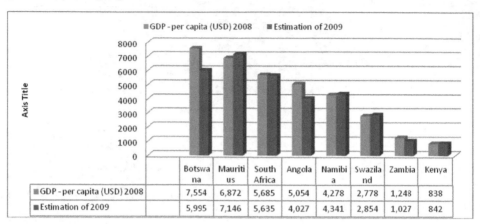

| | Botswana | Mauritius | South Africa | Angola | Namibia | Swaziland | Zambia | Kenya |
|---|---|---|---|---|---|---|---|---|
| ■ GDP - per capita (USD) 2008 | 7,554 | 6,872 | 5,685 | 5,054 | 4,278 | 2,778 | 1,248 | 838 |
| ■ Estimation of 2009 | 5,995 | 7,146 | 5,635 | 4,027 | 4,341 | 2,854 | 1,027 | 842 |

Fig. 7. GDP per person for the period 2008-2009 top 8 countries

Similarly, the last 10 countries with respect to GDP values are shown on Figure 8 below. The leading country in this group is Lesotho whilst Burundi is the last one.

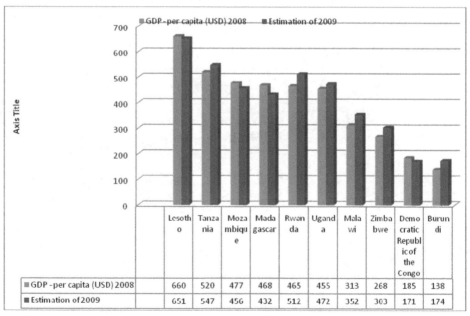

| | Lesotho | Tanzania | Mozambique | Madagascar | Rwanda | Uganda | Malawi | Zimbabwe | Democratic Republic of the Congo | Burundi |
|---|---|---|---|---|---|---|---|---|---|---|
| ■ GDP - per capita (USD) 2008 | 660 | 520 | 477 | 468 | 465 | 455 | 313 | 268 | 185 | 138 |
| ■ Estimation of 2009 | 651 | 547 | 456 | 432 | 512 | 472 | 352 | 303 | 171 | 174 |

Source: World Economic Outlook Database for October 2009

Fig. 8. GDP per person for the period 2008-2009 lowest 10 countries

The ratio of GDP per person/nominal GDP per capita was analysed for the 18 African countries for the period 2008-2009 and results are shown on Figure 9 below.

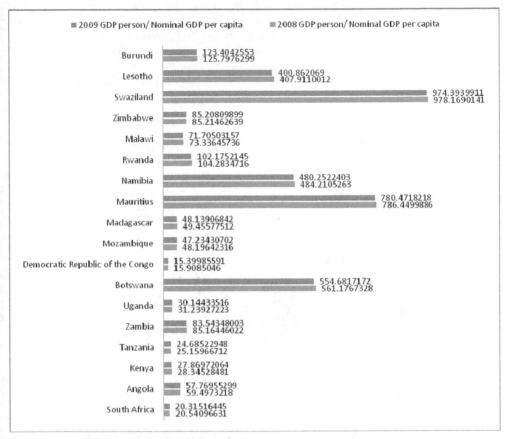

Fig. 9. GDP per person/nominal GDP per capita

The strongest economies showing economic growth are South Africa, Angola, Kenya, Tanzania, Zambia, Uganda, Botswana, DRC, Mozambique, Madagascar, Mauritius, Namibia, Rwanda, Malawi, Zimbabwe, Swaziland, Lesotho and Burundi in that order.

## 4.2 Infodensity for Southern and East Africa countries

There is a close correlation between the country's infostate and GDP per capita. For every point increase in infodensity, GDP per capita increases by an approximate US$150, rendering widespread, affordable access to information services an absolute imperative. The last decade has seen continual growth in infrastructure development and service uptake. Over the last five years, the ITU reports developed and developing countries have increased ICT levels by more than 30%. However, notwithstanding the rapid expansion, to date access and adoption of Internet services is highly unequal across and within countries. Emerging countries face considerable challenges in broad-basing Internet utilization for their growth and development on account of inadequate fixed-line infrastructure, and lack of supporting infrastructure, including electricity and steep prices of personal computers. An approximate 75% of the world populace, a large segment of which lives in emerging markets, consequentially have limited or no access to the Internet. Data on Infodensity was obtained from the International Telecommunications Union (ITU, 2010) and the analysis is shown below on figures 10-14. The fixed teledensity by continent is shown on Figure 10 whilst the mobile cellular subscriptions (%) is shown on Figure 19. Both charts show that Africa has the lowest penetration ratio for fixed teledensity and mobile cellular subscriptions, respectively.

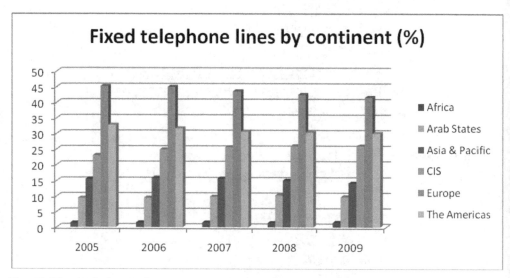

Fig. 10. Fixed teledensity by continent (%)

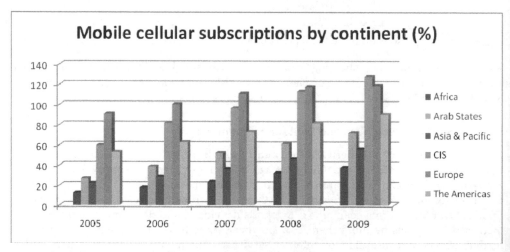

Fig. 11. Mobile cellular subscriptions by continent (%)

The mobile broadband subscriptions by continent are shown on Figure 12 below. Europe leads all other continents whilst Africa remains the least, with at most 2%. The same picture is reflected on internet users shown on Figure 13. The fixed broadband subscriptions shown on Figure 14 show even a wider gap between Africa and other continents.

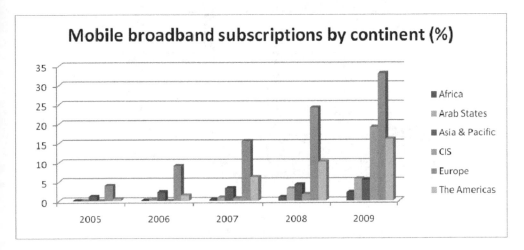

Fig. 12. Mobile broadband subscriptions (%)

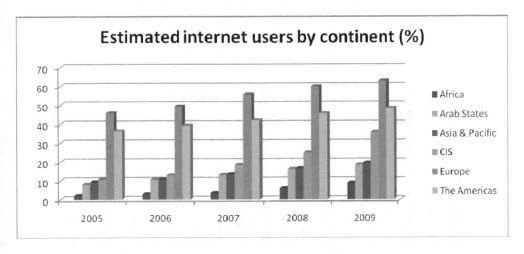

Fig. 13. Internet users by continent (%)

The ICT indicators by country for the 18 East and Southern African countries are shown on Figure 15. South Africa has the highest number of mobile subscribers followed by Botswana and Mauritius. However, Mauritius has the highest density for main telephone lines and internet users.

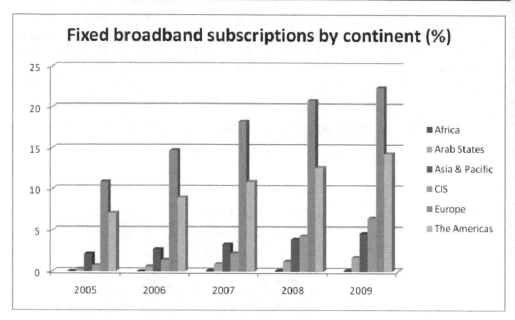

Fig. 14. Fixed broadband subscriptions by continent (%)

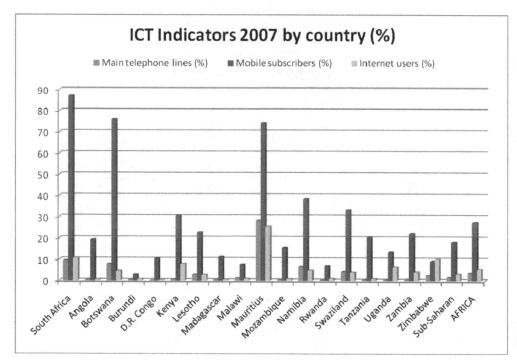

Fig. 15. East and Southern Africa ICT Indicators for 2007

A summary of the statistics on Infodensity as measured by the ICT indicators and the GDP per capita for the 18 countries in East and Southern Africa is shown on Table 2 below. The mean for the 18 East and Southern African countries with respect to main telephone density is 3.8%, mobile subscribers is 27.87%, and internet use is at 4.87%.

|  | N | Minimum | Maximum | Mean | Std. Deviation |
|---|---|---|---|---|---|
| Main telephone lines (%) | 18 | 0% | 28% | 3.80% | 6.791% |
| Mobile subscribers (%) | 18 | 3% | 87% | 27.87% | 25.490% |
| Internet users (%) | 18 | 0% | 25% | 4.87% | 6.124% |
| GDP per capita 2008 (US$) | 18 | 138 | 7,554 | 2,125.33 | 2,554.349 |
| GDP per capita 2009 (US$) | 18 | $171 | $7,146 | $1,996.50 | $2,349.695 |
| Valid N (listwise) | 18 | | | | |

Table 2. Descriptive Statistics for East and Southern Africa

A one-sample T-test for the same data for the 18 East and Southern African countries is shown on Table 3 below. The 95% confidence interval for the lower and upper levels are shown on the same Table 3.

|  | Test Value = 0 | | | | | |
|---|---|---|---|---|---|---|
|  |  |  |  |  | 95% Confidence Interval of the Difference | |
|  | t | df | Sig. (2-tailed) | Mean Difference | Lower | Upper |
| Main telephone lines (%) | 2.374 | 17 | .030 | 3.800% | .42% | 7.18% |
| Mobile subscribers (%) | 4.638 | 17 | .000 | 27.866% | 15.19% | 40.54% |
| Internet users (%) | 3.371 | 17 | .004 | 4.866% | 1.82% | 7.91% |
| GDP per capita 2008 (US$) | 3.530 | 17 | .003 | 2,125.333 | 855.09 | 3,395.58 |
| GDP per capita 2009 (US$) | 3.605 | 17 | .002 | $1,996.500 | $828.02 | $3,164.98 |

Table 3. One-Sample Test for East and Southern Africa

## Correlation between ICT indicators and GDP per capita

The correlation coefficients between the ICT indicators and the GDP per capita for both 2008 and 2009 are summarized in the table 4 below. The correlation coefficients between the main telephone lines (%) to the GDP per capita is 0.721 and 0.798 for the years 2008 and 2009, respectively. The mobile subscriber rate (%) is strongly correlated to the GDP per capita, showing values of 0.881 and 0.902 for the years 2008 and 2009, respectively. However, the correlation coefficients between internet use (%) and GDP per capita remains as low as 0.531 and 0.619 for the years 2008 and 2009, respectively.

| | | GDP per capita 2008 (US$) | GDP per capita 2009 (US$) |
|---|---|---|---|
| Main telephone lines (%) | Pearson Correlation | .721** | .798** |
| | Sig. (2-tailed) | .001 | .000 |
| | N | 18 | 18 |
| Mobile subscribers (%) | Pearson Correlation | .881** | .902** |
| | Sig. (2-tailed) | .000 | .000 |
| | N | 18 | 18 |
| Internet users (%) | Pearson Correlation | .531* | .619** |
| | Sig. (2-tailed) | .023 | .006 |
| | N | 18 | 18 |

Table 4. Correlations between ICT Indicators and GDP per capita

The correlation cofficients between the ICT indicators and the GDP per capita are shown on Figure 16 below.

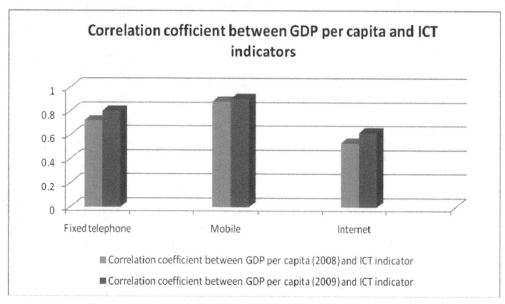

Fig. 16. Correlation between ICT indicators and GDP per capita.

From the above chart on Figure 16, the correlation coefficient between GDP per capita and ICT indicators is highest with the mobile density (about 90%) and then followed by fixed telephony (about 75%), and lowest with internet penetration ratio (about 57%). Hence, the mobile density of a country in East and Southern Africa is a good measure of the relative proportion with the GDP per capita.

## 4.3 Sustainable technology capacity in East & Southern Africa

The ICT policy formulation process went through the due process in all the countries in East and Southern Africa, and its implementation is at various stages from country to country depending the availability of financial resources. Notable achievement have been achieved in ICT Policies, coordination with National ICT Committees, improvement of regional connectivity through coordination ministries, and access to information through the website, portals, SMEs, etc. ICT education and training is required to address the various e-skills opportunities identified in the public sectors. However, social capital/networks have not received much attention with respect to technological development and much work is required to establish basic e-business framework in all countries outside South Africa. Availability of electricity in remote parts of member countries is adversely affecting rapid implementation of the telecommunications infrastructure. The convergence in ICTs is shifting the focus to infrastructure, protocols, applications and services (ERPs), and content as specific areas that need capacity building initiatives.

ICTs affect all the MDGs. There is a strong correlation between ICTs and economic growth. The data collection techniques used in this study include formal meetings, oral interviews, questionnaires and records observation. The major components of the data collected included the identity of the institution; current capacity of the institution; current and future interventions on staff capacity development; the availability and status of the national/corporate ICT policy framework, and progress towards its implementation; quality assurance philosophy and framework; and strategic programme development, management and reviews. The capacity needs of the national and regional institutions/organisations identified by the research and shown on Table 5 below. The research showed evidence of technological capacity with tremendous opportunities that can change the face of East and Southern Africa through identified public institutions in each of the countries covered by the research project. Table 5 below shows the ICT Capacity Needs Components by country, identified by the research.

All the six (6) African countries covered by the qualitative research have National ICT Policy Frameworks which are in different phases of implementation due to challenges in the availability of financial resources. The common strategic areas of focus for meaningful ICTs development in East and Southern Africa are summarised in Table 6 below and these form a basis for the goals and specific objectives for implementation (Government of Zimbabwe Ministry of ICT Strategic Plan 2010-2014, 2010).

It is envisaged that the above key areas for ICTs development would change the landscape for sustainable ICTs development in East and Southern Africa within a period of about five years, as the government support and commitment is very high in all the six African countries visited. The political will and common singleness of heart throughout the public institutions visited needs to be maintained and developed further with the required resources for ease of implementation.

Human capital development is central to capacity development, as adequate human resources are an essential component of a nation's ability to carry out its mission. In pursuit of the broader objectives of the African Union to accelerate economic integration of the continent, with the aim to achieve economic growth, reduce poverty and attain sustainable economic development, the Tripartite Summit of the Heads of State and Government of the

| Capacity Components | Country Regional / National Institution |
|---|---|
| *Policy Formulation and Planning* | Kenya, South Africa, Tanzania, and Zimbabwe |
| *Harmonisation of ICT Infrastructure and facilities* | All countries in East and Southern Africa |
| *Legislative Framework* | East African Community , COMESA, and SADC with support from Government Ministries in All Countries concerned. |
| *Establishing Centres of Excellence*<br>• *Centres of Specialisation*<br>• *Internationalisation*<br>• *Professorial Chairs*<br>• *Adjunct Professors* | South Africa, Tanzania, and Zimbabwe through the Universities/Colleges |
| *Human Capital Development and Institutional Capacity:*<br>• *Staff Development Programmes,*<br>• *Professional Continuous Development,*<br>• *Workshops /Seminars/ Conferences*<br>• *Digital Libraries.* | All Countries in East and Southern Africa through the Universities |
| *E-Skills Development & ICT Training*<br>• Short-term<br>• Long-term<br>• Open and distance learning<br>• Professional Continuous Development | Government Ministries and Universities in All Countries in Southern Africa |
| *Collaborative framework & Research Leadership*<br>• Collaborative Networks<br>• Research Leadership Training/Mentoring | South Africa, Botswana and Tanzania |

Table 5. ICT Capacity Needs Components by Country

| KEY AREA FOR ICT DEVELOPMENT |
|---|
| 1. **Infrastructure establishment and development**, e.g. connectivity, fibre, VSAT, wireless, wireline, VoIP, etc. |
| 2. **Human Capital and Social Networks Development** (*Humanware*), e.g. advocacy, skills, e-literacy, sustainable capacity building, languages, curricula, etc. |
| 3. **Governance,** e.g. policy frameworks, ICT Bill, regulatory framework, corporate governance, etc. |
| 4. **E-government & e-business** e.g. Govt portal, e-commerce frameworks, e-learning, national payment systems, etc. |
| 5. **Application development**, e.g. innovation, animation, e-development, etc. |
| 6. **ICT Industry, Investment, & partnerships**, e.g. PPPs, innovative SMEs, tax incentives, etc. |
| 7. **Research and development**, e.g. Research, Cross and multidisciplinary collaborative projects, etc. |
| 8. **Security and quality assurance frameworks**, e.g. interoperability, quality of service, etc. |
| 9. **Corporate Services**, e.g. internal ministry support requirements, resource mobilisation, etc. |

Table 6. Key Areas for ICTs Development

Common Market for East and Southern Africa (COMESA), East African Community (EAC) and the Southern Africa Development Community (SADC) met in Kampala, Uganda on 22nd October, 2008. The East African Community Partner States have various advantages in terms of geographical proximity, common socio cultural characteristics, and economic complementarities for maximizing the benefits from regional cooperation in the ICT sector. Regional cooperation in ICT does not only facilitate greater access to ICT infrastructure, but is also essential to promote trade, governance and ICT business opportunities within and beyond the region. Regional ICT Special Projects covers COMESA, IGART, etc.

ICTs have tremendous roles in achieving the regional economic integration objectives of SADC, COMESA and EAC (http://www.eac.int/treaty.htm), which include the following:

- Promotion of sustainable growth and equitable development of Partner States including the rational utilization of the region's natural resources and protection of the environment;
- Strengthening and consolidating the long standing political, economic, social, cultural and traditional ties between Partner States and associations between the people of the region in promoting a people-centres mutual development;
- Enhancing and strengthening participation of the private sector and the civil society;
- Mainstreaming gender in all its programmes and enhancement of the role of women in development;
- promoting good governance including adherence to the principle of democratic rule of law, accountability, transparency, social justices, equal opportunities and gender equality; and
- Promotion of peace and stability within the region, and good neighborliness among partners States.

## 5. Conclusion

ICTs impact all the MDGs, especially in eradicating extreme poverty and hunger. The solution to poverty and hunger is not money but knowledge, hence the thrust on human capital development national programmes with a bias towards sustainable social networks/capital to ensure empowerment of local communities and indigenous people. Revolutionary science and technology innovation drive at the lowest level of education, e.g. pre-school, up to universities and colleges and then across all communities is inevitably very critical in the eradication of extreme poverty and hunger. Curriculum reviews for schools and universities to contextualise the technology diffusion and innovation to an African environment require urgent attention. Furthermore, the harmonisation of the infrastructure and equipment facilities for schools, colleges, and institutions that drive education for sustainable development, is equally important. ICTs are therefore key enablers to the generation and dissemination of knowledge, hence the achievement of the MDGs. In fact, ICTs contribute to economic growth, as evidenced by the strong correlation between the GDP growth and ICT indicators. ICTs increase productivity through:

- Better communication and networking at lower costs
- Digitalisation of production and distribution
- New trade opportunities through e-commerce
- Access to knowledge
- Increased competition

The mean for the 18 East and Southern African countries with respect to main telephone density is 3.8%, mobile subscribers is 27.87%, and internet use is at 4.87%. With the exception of South Africa, all the East and Southern African countries are among the late majority and laggards with respect to diffusion and adoption of ICT innovations, i.e. they are largely late end-consumers of technology that has been tried and tested from developed and some developing countries. The ICT capital comprises network infrastructure and ICT machinery and equipment. ICT and non-ICT factor inputs are combined to produce ICT and non-ICT goods and services, without a one-to-one correspondence. There is a strong correlation between ICT diffusion and high economic growth. The correlation coefficient between GDP per capita and ICT indicators is highest with the mobile density (about 90%) and then followed by fixed telephony (about 75%), and lowest with internet penetration ratio (about 57%). Hence, the mobile density of a country in East and Southern Africa is a good measure of the relative proportion with the GDP per capita.

The methodology used was largely qualitative on technology capacity needs assessment that covered 6 countries, and also quantitative on GDP and Infodensity covering 18 countries in East and Southern Africa. GDP and Infodensity data was collected for 18 African countries to ascertain the link between ICTs diffusion and GDP density per country. Policy-formulation frameworks and sustainable capacity building provide a conducive environment for meaningful development in the SADC countries. Capacity needs assessment included both the human capital development and social capital aspects in order to achieve sustainable information and communication technology capacity development. Human capital development is central to capacity needs. There is a strong correlation between ICT diffusion and high economic growth. The solution to poverty and under-development in African countries is knowledge and economic empowerment. The recommended sustainable technology development with an African model is proposed.

The East and Southern African countries covered by the study showed tremendous development potential, even though they are among the late majority and laggards with respect to technological innovations. The recommended sustainable technology development African model is proposed with the following major components:

1. Human capital development national programmes with a bias towards social networks/capital to ensure empowerment of local communities and indigenous people
2. Curriculum reviews for schools and universities to contextualise the technology diffusion and innovation to an African environment
3. Revolutionary science and technology innovation drive at the lowest level of education, e.g. pre-school, up to universities and colleges and then across all communities.
4. Universitisation of the entire education system where the academic leadership offered by academia and research institutions inculcates and influences the curricular of the entire education system in order to provide a meaningful contribution to knowledge
5. Harmonisation of the infrastructure and equipment facilities for schools, colleges, and institutions that drive education for sustainable development
6. Review of the legal framework and policy formulation mechanism with a view to rapid development initiatives
7. Establishing centres of excellence in specialised fields to provide leadership (academic, research and consultancy) on key developmental issues.

# 6. References

Baliamoune-Lutz Mina (2003), *An analysis of the determinants and effects of ICT diffusion in developing countries*, Information Technology for Development 10 (2003) 151–169 151, IOS Press.

Government of Zimbabwe (2010), *"The Ministry of Information Communication Technology (MICT) Strategic Plan 2010-2014"*, Republic of Zimbabwe Government Printers, February ,2010.

Heeks, R. (1999) 'Information and Communication Technologies, Poverty and Development'. Development Informatics Working Paper Series, Paper No. 5, June 1999, IDPM, Manchester.
    http://www.man.ac.uk/idpm/idpm_dp.htm#devinf_wp.

International Telecommunications Union (2010), World Telecommunication/ICT Indicators Database: ITU, June, 2010,
    http://www.itu.int/ITU-D/ict/statistics/index.html.

Kabanda, G. (2008), *Collaborative opportunities for ICTs development in a challenged African environment*, International Journal of Technology Management & Innovation, Volume 3, Issue 3 of 2008, pages 91-99, ISSN 0718-2724.

Kekana, N. (2000) Information Communication and Transformation: A South African Perspective. Communication, Vol. 28 No.2, p54-61

Kyem, P.A. and LeMaire, P.K. (2006), "Transforming recent gains in the digital divide into digital opportunities: Africa and the boom in mobile phone subscription", The Electronic Journal on Information Systems in Developing Countries, EJISDC (2006) 28, Volume 5, pages 1-16.

Moradi, M.A. and Kebryaee, M. (2010), "Impact of Information and Communication Technology on Economic Growth in Selected Islamic Countries",
    http://www.ecomod.org/files/papers/987.p.

Rogers, E.M. (1995). Diffusion of innovations (4th edition). New York: The Free Press.

Rogers, E.M. (2003). Diffusion of Innovations (5th Edition). New York. The Free Press.

Samiullah, Y and S. Rao (2000) 'Role of ICTs in Urban and Rural Poverty Reduction'. Paper prepared for MoEF-TERI-UNEP Regional Workshop for Asia and Pacific on ICT and Environment, May 2000, Delhi.
    http://www.teri.res.in/icteap/present/session4/sami.doc

Sciadas G. (2003), "Monitoring the Digital Divide and beyond", The Orbicom project publication, in collaboration with the Canadian International Development Agency, the *Info*Dev Programme of the World Bank and UNESCO: Claude-Yves Charron, Canada, 2003 (ISBN 2-922651-03-7), page 10.
    http://www.orbicom.uqam.ca.

UNCTAD (2006), The United Nations Conference on Trade and Development "The least developed countries report for 2006: developing productive capacities", Paper number UNCTAD/LDC/2006, United Nations New York and Geneva, 2006 (http://www.unctad.org/en/docs/ldc2006_en.pdf).

UNDP. (2001). Information communications technology for development. Essentials: Syntheses of Lessons Learned No. 5. [Online]. Retrieved 28 September, 2006 from: http://www.undp.org/eo/documents/essentials_5.PDF

UNESCO (2010), "Universities and the Millennium Development Goals Report 2007", Association of Universities and Colleges in Canada, WDI publication, presented at the Association of Commonwealth Universities conference, Cape Town, April 2010.

World Development Report (1998/99). The power and reach of Knowledge. [Online]. Retrieved 15 January, 2006 from:
http://www.worldbank.org/wdr/wdr98/ch01.pdf Accessed 15/01/06

World Bank (2003). *Lifelong learning in the global knowledge economy: Challenges for developing countries.* Washington, DC: World Bank.

World Bank (2002). *ICT Sector Strategy Paper.* World Bank.
http://info.worldbank.org/ict/ICT_ssp.html

World Development Indicators (2004). "E-Strategies: Monitoring and Evaluation Toolkit", The World Bank INF/GICT Volume 6.1B 3rd January, 2005.
http://www.worldbank.org/ict/.

# 10

# Towards the European Union – Impact of FDI and Technological Change on Turkish Banking

Özlem Olgu

*College of Administrative Sciences and Economics, Koc University,*
*Rumeli Feneri Yolu, Sariyer, Istanbul*
*Turkey*

## 1. Introduction

Despite the fact that Turkey has the 17th largest economy in the world and the 6th largest economy in the European Union (EU), its financial system is still relatively small compared to most of the EU countries (State Planning Organisation, 2009). Even though increased competition has been monitored over the last decade particularly due to removal of barriers on foreign entry into the Turkish banking sector, there are still doubts on Turkey's competitive strength with the EU countries. What is important is that financial performance of institutions has become a primary concern of investors, lenders, shareholders, and in particular to managers, in planning and controlling their activities. We believe, financial progress since the 2000-01 twin crises, enforcement of internationally accepted banking regulatory and supervisory standards, for instance Basel I and II, as well as implementation of macroeconomic reforms increased stability and trust in Turkey. We expect these developments to have positive impact on bank performances attracting foreign direct investment (FDI) to the banking sector and as a result increase transfer of know how and apllication of new technological advances.

Above all the most imperative occurrence, not only for Turkish banking but also for Turkey as a nation, is the declaration of it as an EU accession country in 2005. After more than four decades of long standing negotiations, Turkey finally became a candidate member of the EU and thus required to adjust its legislation and regulatory environment to that of the EU by adapting the EU acquis communautaire during the convergence process.[1] The current government carried out a constitutional revolution: deepening and widening democratic freedoms, introducing minority rights for the Kurds and, above all, starting to subordinate

---

[1] One of the most important items in the acquis on financial institutions is the harmonization of capital adequacy regulations. The latest EU directive that sets out the capital adequacy framework is the Capital Adequacy Directive 3 (CAD 3). CAD 3, which took effect in 2007 and applied to all credit institutions in the EU, simply translates Basel II into EU legislation. The adaptation of a Basel II type regulatory framework for financial institutions by the EU facilitated Turkey's efforts to comply with this section of the acquis, as Turkish authorities already opt to implement Basel II provisions fully as of January 2008 for capital adequacy requirements. In fact, capital adequacy requirements in Turkey were already established inline with Basel I (adopted in 1989) and market risk was incorporated in the calculation of capital adequacy ratio in 2002.

Turkey's army to civilian authority. In fact, the European project worked as a powerful engine of reform and helped glue together Turkey's political tribes. We believe the ultimate hope of EU membership is an important factor on augmented FDI and would make Turkey even more attractive for foreign investors in the near future. This is due to the fact that Turkey has a highly skilled and adaptable labor force, a large domestic market, and the advantage of geographic closeness to Europe, Middle East, Northern Africa, and Central Asia markets.

The Global Competitiveness Report 2007-2008 released by the World Economic Form listed Turkey in the 53rd place among 131 countries in the overall ranking (six step going up from 2006). Moreover, Turkey is listed as the 13th most attractive country in the world for FDI, thus it is not surprising to have approximately 53 per cent of total banking assets held by foreign investors (BRSA, 2007). Since 2005, numerous leading banks in the global arena, particularly with European origin, raised their shareholdings in large and medium sized Turkish commercial banks. For instance, 42 per cent stake of Turkiye Ekonomi Bankası (TEB) bought by France's BNP Paribas in 2005, followed by 89 per cent stake acquisition of Dışbank by Belgium's Fortis bank, 5.5 per cent share sale of Garanti Bankası to the US General Electric Consumer Finance, and 50 per cent stake sale of Yapı Kredi Bankası to Italy's UniCredito.

Even though massive progress has been achieved in terms of foreign inflows over the recent years, Turkey still ranks as one of the bottom compared to the EU emerging countries. The 28.7 percent average share of foreign participation in the EU-27 as well as Bulgarian (93 per cent) and Romanian (94 per cent) banking sectors is well above the level in Turkey (24.8 per cent). This is supportive of the argument by Steinherr et al. (2004) that Eastern European countries did not have any historical commercial banking activity when they opened up their markets and the most efficient option of quickly transforming their banking system was to invite foreign banks as strategic investors. In contrast, Turkish banking sector has long been established with domestic banks and provides the foreign banks with one representative office in order to help manage their activities. Additionally, legal uncertainties, family ownership of private banks, and relationship lending practices in Turkey could be other concerns of foreign investors over the past decades.

This chapter contributes to the literature by implying that technology transfer associated with inward FDI yields higher bank productivity. More generally, analysing the dynamic effects of foreign presence may be important in seeking answers to some of the relevant questions about effects of FDI on host country banking industries and technology transfers. It is interesting to illustrate that even though research on transition economies has been growing, studies on Turkish banking sector lag far behind. None of the studies revised in the literature identified productivity differences among foreign invested and domestic banks in Turkey as well as not examining the most recent periods. What distinguishes this study from the rest in the literature is its comphrehensive analysis of issues specified in various individual studies. The literature is extended in this study by analysing the impact of foreign ownership on the DEA Malmquist productivity scores and its decomposed components. Hence, informtion on bank technical efficiency change (TEC), pure efficiency change (PEC), technological change (TC), scale efficiency change (SEC) as well as total factor productivity change (TFPC) over the analysis period would be valueable and may serve to inform the relative effect of increased FDI in the market as Turkey is considered as a potential EU member.

This study focuses on 17 commercial banks operating in Turkey from which more than half of them received foreign investment since 2001. The 'first-stage' analysis has been introduced with an application of the input oriented DEA Malmquist index. The 'second-stage' analysis tests the impact of bank specific variables as well as several dummy vaiables on the decomposed DEA Malmquist index components following the studies by Delis *et al.* (2008), Mukerjee *et al.* (2001), Dogan & Fauesten (2003), Pasiouras *et al.* (2007), and Lensink *et al.* (2008). Regarding foreign bank entry, we expect to find evidence that productivity of domestic banks have increased particularly due to developments in application of technological advances.

In the next section, long standing negotiations between the EU and Turkey as well as the probable outcome of Turkey' EU membership both in politcal and economic grounds is summarised. In section 3, a summary of relevant studies in the literature is introduced followed by description of the data in section 4. Section 5 characterizes firstly the methodology of decomposing DEA-Malmquist productivity measures into its relevant components and secondly how the estimated bank TFP change scores as well its decomposed components are related to relevant bank specific factors in the second stage regressions. Section 6 summarises the empirical findings followed by conclusions in section 7.

## 2. Long-standing negotiations and impact of Turkey's probable EU membership

The EU has gone through a massive enlargement process in 2004 by inclusion of ten new countries in its body followed by Romania and Bulgaria in 2007. Turkey attached particular importance to the EU's recent enlargement process for two main reasons: (i) playing an active role in the ex-Soviet bloc as a trade partner, and (ii) the Association Agreement between Turkey and the EU guarantees its full EU membership. In fact, Turkey was one of the first candidate countries applied to join the European Economic Community (EEC), in July 1959, shortly after its creation in 1958. The response from the EEC was a suggestion to establish an Association between the Republic of Turkey and the EEC, i.e. the Ankara Agreement (1963), until circumstances in Turkey permit its accession.[2]

After pursuing inward-oriented development strategies throughout the 1960s and 1970s, Turkey switched over to a more outward-oriented policy stance in 1980 aiming to integrate the country into the EU. Turkey applied for full membership in 1987, whilst received a response in 1990 stating that accession negotiations could not be undertaken at the time as the EU was engaged in major internal changes (adoption of the Single Market), and developments in Eastern Europe and the Soviet Union. However, the EU was prepared to extend and deepen economic relations with Turkey without rejecting the possibility of full membership at a future date and the Commission underpinned the need for a comprehensive cooperation program to facilitate integration of the two sides as well as finalizing the Customs Union by 1995.

---

[2] The Ankara Agreement aimed at securing Turkey's full membership in the EEC through the establishment of a customs union in economic and trade mattters in three phases, which still constitutes the legal basis of the Association.

At the Association Council of 29 April 1997, the EU reconfirmed Turkey's eligibility for membership and asked the Commission to prepare recommendations to deepen the Turkey-EU relations, while claiming that the development of this relationship depends on a number of issues relating to Greece, Cyprus and human rights. The Commission, however, excluded Turkey from the enlargement process in its report entitled "Agenda 2000" disclosed on 16 July 1997, which may be seen as a contradiction. The report granted that the Customs Union was functioning satisfactorily and that it had demonstrated Turkey's ability to adapt to the EU norms in many areas, whilst porposed a number of recommendations on liberalization of trade in services, consumer protection and a number of political issues as pre-conditions for moving the relations forward.

On December 10–11, 1999, the European Council meeting held in Helsinki produced a breakthrough in Turkey-EU relations, where Turkey was officially recognized as a candidate country for accession and signed an Accession Partnership with the EU. However, contrary to other candidate countries (EU-10[3], Bulgaria and Romania), Turkey did not receive a timetable for accession. After the approval of the Accession Partnership by the Council and adoption of the Framework Regulation on February 26, 2001, the Turkish government announced on March 19, 2001, its own National Program for adoption of the *acquis communautaire*. Since then, progress toward accession continues along the path set by the National Program. At the December 2004 Cophenhagen European Council meeting, it was decided to launch negotiations with Turkey and establish a timetable for accession (European Commission, 2004b).

Turkey took a number of important steps towards this end, such as major review of the Turkish Constitution that thirty-four articles have recently been amended. The package of constitutional amendments covers a wide range of issues, such as improving human rights, strengthening the rule of law and restructuring of democratic institutions. In addition, numerous reform measures have been adopted in economic framework in line with the National Program. Even though the membership negotiations has been opened, great uncertainties continue to prevail about whether Turkey will be able to achieve its goal of accession to the EU. Some of these are in political grounds whilst the greatest uncertainty might be whether EU governments and societies are willing to accept a large but nonetheless a Muslim country as part of the EU.

In the case of EU membership, massive population of Turkey will make it the second largest country of the EU following Germany, whilst the population projections state that it will likely be the largest by 2025. Togther with Germany, they will account for almost 30 percent of the EU's population and the EU will then have seven large countries, that is, Germany, France, Italy, the UK, Poland, Spain and Turkey.

Referring to its massive population, Turkey will certainly impact strongly on the political dynamics both among large members and within the EU as a whole. At this point, we find it interesting to examine the likely direct impact of Turkey's EU membership on the European Council, Commision and Parliament. Table 2 sets out voting weights by population share in an EU-25, EU-27 and probable EU-28.

---

[3] EU-10 countries are; Slovenia, Slovakia, Malta, Cyprus, Latvia, Lithuania, the Czech Republic, Estaonia, Poland and Hungary.

| Countries | 2003 | 2015 | 2025 | 2050 |
|---|---|---|---|---|
| France | 60144 | 62841 | 64165 | 64230 |
| Germany | 82476 | 82497 | 81959 | 79145 |
| Italy | 57423 | 55507 | 52939 | 44875 |
| Poland | 38587 | 38173 | 37337 | 33004 |
| Spain | 41060 | 41167 | 40369 | 37336 |
| UK | 59251 | 61275 | 63287 | 66166 |
| Turkey | 71325 | 82150 | 88995 | 97759 |
| Total EU-25 | 454187 | 456876 | 454422 | 431241 |
| Total EU-27 | 484418 | 485692 | 481837 | 454559 |
| Total EU-28 (incl Turkey) | 555743 | 567842 | 570832 | 552318 |
| Turkey as % of EU-28 | 12.8 | 14.4 | 15.5 | 17.7 |

Source: Adapted by the author from UN population forecasts

Table 1. Population projections for Turkey and selective EU countries, 2003-2050

| Countries | Share in EU-25, 2004 | | Share in EU-27, 2007 | | Share in EU-28, 2015 | |
|---|---|---|---|---|---|---|
| | V.W | Seats | V.W | Seats | V.W | Seats |
| Germany | 18.1 | 99 | 16.9 | 98 | 14.5 | 82 |
| France | 13.2 | 78 | 12.9 | 78 | 11 | 64 |
| Italy | 12.6 | 78 | 11.4 | 78 | 9.7 | 64 |
| UK | 13 | 78 | 12.6 | 78 | 10.7 | 64 |
| Spain | 9 | 54 | 8.4 | 54 | 7.2 | 44 |
| Poland | 8.4 | 54 | 7.8 | 54 | 6.7 | 44 |
| Turkey | -- | -- | -- | -- | 14.4 | 82 |

Source: Adapted by the author from UN World Population Division, World Population Prospects (2002)
-V.W: Voting weights

Table 2. Voting weights and number of seats in the European Parliament

Assuming the EU agrees on 'double-majority' system of voting in the constitution, EU decisions will need a majority of two countries and population. In an EU-28, no proposal could be passed without the support of at least 15 member countries. In such a system, no single state can dominate; where population size has more power is through the ability to block decisions. If the threshold is set at 60 percent, then in the EU-25, Germany together with the UK and France can block decisions (with 44.3 percent of total population, or with Italy instead of the UK making 43.9 percent), though they cannot achieve this in an EU-28 (where they have 36.2 percent of population) unless the population majority is set to 65 percent (Hughes, 2004).

In the case of a EU-28, each of Turkey and Germany will have around 14.5 per cent of the vote. Even though they will be strong players, they can not block proposals even together but with a third large country. The largest 5 countries in an EU-28 will account for 60.3 per cent of the vote by population. This is only 3.4 per cent higher than the share of the 'big 4' countries in the EU-25 (where they have 56.9 per cent of the vote). So, Turkey will be an important powerful player and will add to the already complex set of alliances and blocking combinations that are possible. But in an EU-28, despite its size, it does not add strongly to the dominance of large countries. As the debate on voting power in the constitution shows, questions of power and votes are highly politically sensitive and negotiations for Turkey's accession will not be simple.

Moreover, given its size, Turkey will have a large impact on the European Parliament. Assuming the EU decides to keep a limit of 732 seats in the parliament, then all countries' allocations have to be reduced to avoid the accession of Turkey, Romania and Bulgaria adding 149 seats (99 (same as Germany), 33 and 17, respectively). If there was a simple proportionate reduction across all countries on the allocation of seats, Germany and Turkey would have 82 seats and 11.2 per cent share of the vote each (down from 13.5 per cent), France, Italy and the UK would have 64 seats representing 8.7 per cent of the vote each (down from 10.6 per cent), and Spain and Poland would have 44 seats and 6.0 per cent of the vote (down from 7.3 per cent).

From the economic perspective, the most important fact about Turkey is its low purchasing power parity (PPP) adjusted per capita income which is almost equal to those of Bulgaria and Romania but more complicated, and expected to create economic consequences for the EU. What is different from the current EU is that its largest members today – Germany, France, Italy and the UK – also have the largest economies. Political and economic dominance go together. However, this is not the case for Turkey. Moreover, Turkey's biggest problems are its high unemployment and inflation rates as well as current account deficit. But we should emphasize the high GDP growth and GDP per capita growth rates that can be accepted as indicators of good performance and promise of development in the near future.

From Turkey's perspective, the EU accession will grant numerous benefits. Actually, Turkey would not have to wait very long to start reaping the benefits of an eventual EU accession. With the opening of EU accession negotiations in October 2005, Turkey is likely to attract larger sums of FDI in the near future, which has already been experienced in the banking sector since 2005. The opening of EU negotiations acted as a strong signal that Turkey will eventually become a full member of the EU and would assure foreign investors that the Turkish economy will follow a stable growth path for the foreseeable future and the legal and judicial environment will improve across all relevant areas of the common acquis.

## 2.1 Why does EU opposes to Turkey's membership?

While the EU is working with Turkey to help it move forward in its probable EU membership, there are some who are concerned about this progress. Some of the main issues pointed out by the opposers can be summarised as follows:

First, Turkey's culture and values are different from those of the EU as a whole. In fact, Turkey's 99.8 percent Muslim population is too different from Christian-based Europe. However, the EU introduces this case in a very political manner and explains that the EU is not a religion-based organization and that 12 million Muslims currently live throughout the EU, Moreover, Turkey is a secular (a non-religion-based government) state, whilst needs to

"Substantially improve respect for the rights of non-Muslim religious communities to meet European standards."

Secondly, another argument is based on the geographical fact that Turkey is mostly not in Europe, thus it should not become part of the EU. If that is the case what could be said about Cyprus? Is the island geographically located in Europe? This brings us to the third issue of Turkey's non-recognition of Republic of Cyprus, which is a full-fledged member of the EU since 2004. Additionally, many are concerned about the rights of Kurds in Turkey. The Kurdish people have limited human rights and there are accounts of genocidal activities that need to stop for Turkey to be considered for the EU membership.

Thirdly, Turkey is receiving considerable assistance from its European neighbors as well as from the EU. The EU has allocated billions and is expected to allocate billions of euros in funding for projects to help invest in a stronger Turkey that may one day become a member of the EU. The low per-capita income of the Turkish population is also of concern since the economy of Turkey as a new EU member might have a negative effect on the EU as a whole. From the EU's perspective, the most probable economic objections to Turkey's full membership can be listed as: (i) Turkey will receive a significant part of the EC structural funds and will impose an additional burden on countries that are major contributors to the Community budget; (ii) the Turkish economy is not mature enough for the single market, and the Turkish industry is not competitive with that of the EU, and (iii) there is also a problem of free movement of the labor force as Turkey will lead to a huge arrival of Turkish immigrants into the EU countries, particularly Germany.

Finally, some are concerned that Turkey's large population would alter the balance of power in the EU as discussed in the previous section of this chapter. After all, Germany's population, which is the largest country in the EU is only at 82 million and declining. Turkey would be the second largest country and perhaps eventually the largest with its much higher growth rate in the EU. Thus it would have considerable influence in the European Parliament.

## 3. Literature survey

Over the last decade several papers examined the relationship between foreign ownership and bank performances as well as the differences among foreign owned versus domestic banks. Even though there is an established literature on identifying the effects of FDI inflows and access to foreign capital on the productivity of non-financial institutions, such as Gorg & Greenaway (2004) and Moran *et al.* (2005); empirical research analysing the impact of increased FDI on bank productivity is very narrow. Studies by Classens *et al.* (2001); Demirguc-Kunt & Huizinga (2000); Berger & Humphey (1997) and Berger *et al.* (2000) examined performance of domestic and foreign-owned banks offering little insight on how foreign existence influences bank productivity. Moreover, several studies in the literature test the impact of foreign ownership on bank efficiency, mainly using ownership dummies (see for instance, Lensink et al., 2008), although none can be found on the foreign ownership-bank productivity relationship.

A channel of the literature works on analysing sources of bank productivity differences across a sample of countries mainly focusing on country-specific economic, demographic and technological conditions (Chaffai *et al*, 2001); financial regulatory environment (Delis *et al*, 2008), success or failure of policy initiatives (Casu *et al.*, 2004), and economic liberalization or financial deregulation (Berg *et al*, 1992; Dogan & Fausten, 2003; Grifell-Tatje & Lovell, 1997; Isik & Hassan, 2003; Mukerjee *et al*, 2001; Tirtiroglu *et al.*, 2005; Worthington, 1999). The general conclusion reached by Dietsch & Lozano-Vivas (2000); Cavallo & Rossi (2002); Lozano-Vivas et al. (2002); Pasiouras *et al.* (2007); Lensink *et al.* (2008) and Pasiouras (2008) is that country-specific environmental conditions and deregulatory policies are important factors on both bank productivity and efficiency.

In connection to the global advantage hypothesis introduced by Berger *et al.* (2000); Aitken & Harrison (1999) proposed FDI as a factor negatively affecting domestic bank productivity. They stated that foreign banks may have a competition advantage due to their better resources and technologies resulting in greater market share with lower interest margins and risk premiums in a country than its domestic counterparts.

Another argument in the literature is on the cherry-picking aspect, i.e. foreign banks may magnify the risk profile of their domestic counterparts by using their financial power to pick the most rewarding features of the domestic market, and force domestic banks to do more risky business. Furthermore, domestic bank competitiveness and efficiency is achieved before rather than after the entry of foreign banks into the sector. In other words, efficiency is a pre-condition rather than a result of foreign bank entry. Berger *et al.* (2000) focused on this issue specifically examining the major industrialised countries and investigated whether the 'home-field advantage hypothesis' or the 'global-advantage hypothesis' holds.

This isuue has been examined by Hymer (1976) in an earlier study and illustrated that foreign firms are likely to face competitive disadvantages relative to national firms. This is due to the fact that domestic firms have the general advantage of better information about their country's economy, language, laws and politics. This leads to the hypothesis that foreign banks suffer more from a bad institutional framework in the host country than domestic banks. Foreigners and nationals may receive different treatment from governments, consumers and suppliers. In countries with a solid institutional framework, the impact of foreign ownership on bank efficiency will be less negative or more positive than in countries where the institutional framework is bad.

In another study, Mian (2006) develops a theoretical model on the effect of institutional distance on foreign bank behaviour. He assumes that institutional distance between the home country and the host country will cause higher informational, agency, or enforcement costs for foreign banks operating abroad. This issue has also been investigated by Stein (2002) who stated that technologies such as derivative contracts, ATM networks and Internet banking allow banks to interact efficiently with customers over long distances as well as improving the ability of senior headquarters managers to monitor junior managers working at distant locations.

In the case of US banking sector, Hancock *et al.* (1999) suggested that large financial institutions dominate electronic payments processing due to information-based scale economies. Moreover, Berger & Mester (2003) confirmed the hypothesis that technological progress allows banks to offer wider varieties of services and banks engaging in merger activity had the greatest gains in profit productivity. This suggests that merger and acquisitions may allow foreign banks an opportunity to apply technological innovations in host countries.

In the case of Turkey, changing economic and financial environment attracted attention of researchers.[4] Oral & Yolalan (1990) computed operating efficiency and profitability of bank branches and demonstrated service-efficient bank branches as the most profitable. In another study, Zaim (1995) evaluated the effects of liberalization policies on efficiency of Turkish commercial banks during the period of 1981 and 1990 and found that financial liberalization had a positive effect on both the technical and the allocative efficiencies, and public banks were more efficient than private counterparts. Similarly, Denizer, Dinc, & Tarimcilar (2000) examined the bank efficiency during the pre and post-liberalization environment and investigated the scale effects on efficiency by ownership covering the 1970-1994 period. Findings suggested that liberalization programs were followed by an observable decline in efficiency and that the Turkish banking system had a serious scale problem due to macroeconomic instability.

Furthermore, Yildirim (2002) analysed efficiency performance of Turkish banks over the 1988-1999 period. The empirical results suggested that over the sample period both pure technical and scale efficiency measures showed a great variation, the sector could not achieve sustained efficiency gains and that the trend in the performance scores suggested a strong impact of macroeconomic conditions on the efficiency measures. Consistent with Denizer et al. (2000), Yildirim concluded that the sector suffered mainly from scale inefficiency due to decreasing return to scale (DRS). Confirming these findings Isik & Hassan (2002) indicated that the dominant source of inefficiency in Turkish banking was due to technical inefficiency rather than allocative inefficiency, which was mainly attributed to diseconomies on scale. The results suggested that the heterogeneous characteristics of banks have significant impact on efficiency.

Yolalan (1996) used financial ratios to analyze the efficiency of the Turkish commercial banks over the period 1988-1995. The results showed that foreign banks were the most efficient group, followed by private and public banks respectively. Mercan, Reisman, Yolalan, & Emel (2003) introduced a financial performance index on Turkish banks over the 1989-99 period, which allowed for observing effects of scale and mode of ownership on bank behaviour. They applied DEA to select fundamental financial ratios for the period of 1989-1999. The results showed that banks that were taken over by a regulatory government agency perform poorly with respect to their DEA performance index values.

## 4. The data

The data for this study comprise the population of 17 commercial banks operating in Turkey and concerns the 1992-2008 period, consisting of 272 observations, as available from the Bank Association of Turkey (BAT). The sample period is chosen as it covers financial structural changes, the most recent wave of FDI and opening up of EU negotiations.

Selection of inputs and outputs is guided by the objectives of the Turkish banking system, where commercial banks act as intermediaries with the objective of collecting deposits. Therefore, we use the Intermediation approach proposed by Sealey & Lindley (1977), similar to many other studies, Zaim (1995), Kraft & Tirtiroglu (1998), Rezvanian & Mehdian (2002),

---

[4] There are couple of studies that analyse efficiency of Turkish banks applying DEA such as Oral and Yolalan (1990), Altunbas et al. (1994c), Zaim (1995), Denizer et al. (2000), and Isik and Hassan (2002).

Isik & Hassan (2002, 2003) and Havrylchyk (2006), where total assets, total deposits and total expenses (inclusing personnel expenses) are considered as inputs used to produce total loans and interest income as outputs. Table 3 introduces the sample characteristics.

| Year | Domestic | Foreign Invested | Total | Observations |
|---|---|---|---|---|
| 1992-1996 | 17 | 0 | 17 | 68 |
| 1996-2000 | 17 | 0 | 17 | 68 |
| 2000-2004 | 15 | 2 | 17 | 68 |
| 2004-2008 | 8 | 9 | 17 | 68 |
| Total | 57 | 11 | 68 | 272 |

| Approach | Outputs | Inputs |
|---|---|---|
| | Total loans | Total assets |
| Intermediation | Interest income | Total expenses |
| | | Total Deposits |

**FDI Information**

| Year | Domestic Bank | Aquirer | Origin of Aquirer | Stake Bought (%) |
|---|---|---|---|---|
| 2001 | Demirbank | HSBC | UK | 100 |
| 2002 | Koc Bank | UniCredit | Italy | 50 |
| 2005 | YKB | UniCredit/Kocbank | Italy-Turkey | 57 |
| | TEB | BNP Paribas | France | 50 |
| | Disbank | Fortis | Belgium | 89 |
| | Garanti Bankasi | GE Capital Corporation | USA | 26 |
| | Finansbank | National Bank of Greece | Greece | 46 |
| 2006 | Denizbank | Dexia Bank | Belgium | 75 |
| | Sekerbank | BTA | Kazakhstan | 34 |
| | Akbank | Citigroup | USA | 20 |

Table 3. Sample characteristics

## 5. Methodology: DEA-Malmquist index

The nonparametric DEA-Malmquist index can be estimated by exploiting the relationship of distance functions to the technical efficiency measures developed by Farrell (1957). This technique is a "primal" index of productivity change. Therefore, it does not require cost or revenue shares to aggregate inputs and outputs, and was introduced to the literature by Caves et al, (1982a). In order to calculate Malmquist index, it is first required to define distance functions with respect to two different time periods, such as:

$$D_o^t\left(\chi^{t+1}, y^{t+1}\right) = \inf\left\{\Theta : \left(\chi^{t+1}, y^{t+1}/\Theta\right)\right\} \in S^t \qquad (1)$$

Where $D_o$ is the distance function at time t with $\chi^t$, Input vector in time t+1 and $y^t$, output vector in time t+1. Technical efficiency is indicated by $\Theta$ and production technology by $S^t$. The distance function in equation (1) measures the maximal proportional change in outputs required to make $\left(\chi^{t+1},y^{t+1}\right)$ feasible in relation to technology at t (Fare, Grosskopf, Norris & Zhang, referred to as FGNZ, 1994). Then, the Malmquist index reference to technology t is defined by CCD (1982a) as:

$$M_{CCD}^t = \frac{D_o^t\left(\chi^{t+1},y^{t+1}\right)}{D_o^t\left(\chi^t,y^t\right)} \tag{2}$$

whereas for period t+1 it is:

$$M_{CCD}^{t+1} = \frac{D_o^{t+1}\left(\chi^{t+1},y^{t+1}\right)}{D_o^{t+1}\left(\chi^t,y^t\right)} \tag{3}$$

In order to avoid choosing "an arbitrary benchmark", the output based Malmquist productivity index, $M_0$, is specified to be the geometric mean of equation (3). Then:

$$M_o\left(\chi^{t+1},y^{t+1},\chi^t,y^t\right) = \left[\frac{D_o^t\left(\chi^{t+1},y^{t+1}\right)}{D_o^t\left(\chi^t,y^t\right)}\right]\left[\frac{D_o^{t+1}\left(\chi^{t+1},y^{t+1}\right)}{D_o^{t+1}\left(\chi^t,y^t\right)}\right]^{\frac{1}{2}} \tag{4}$$

Following FGNZ (1994), the above formula can be decomposed into technical efficiency change and technological change thus:

$$M_o\left(\chi^{t+1},y^{t+1},\chi^t,y^t\right) = \left[\frac{D_o^t\left(\chi^{t+1},y^{t+1}\right)}{D_o^t\left(\chi^t,y^t\right)}\right]\left[\left(\frac{D_o^t\left(\chi^{t+1},y^{t+1}\right)}{D_o^{t+1}\left(\chi^{t+1},y^{t+1}\right)}\right)\left(\frac{D_o^t\left(\chi^t,y^t\right)}{D_o^{t+1}\left(\chi^t,y^t\right)}\right)\right]^{\frac{1}{2}} \tag{5}$$

Technical Efficiency          Technological Change
Change

The first figure in parenthesis measures the efficiency change, whereas the second one represents the technical change component of the index. There is productivity growth if $M_o$ > 0, stagnation if $M_o$ = 0, and productivity decline if $M_o$ <0. Given the fact that the Malmquist decomposition of FGNZ (1994) is based on CRS reference technology, no scale effect could be identified. Thus, FGNZ' catch-up component from the combination of efficiency change and scale efficiency change is inappropriate. In order to overcome this limitation Ray & Desli (1997) developed a model which measures the correct productivity change by the ratio of constant returns to scale (CRS) distance functions even though the technology is variable returns to scale (VRS). The technological change component based on the VRS distance function could affect scale efficiency change while technical change remains unaffected. The scale efficiency change (SE) component is defined by the distance function as:

$$SE^t\left(\chi^t,y^t\right)=\frac{D_{CRS}^t\left(\chi^t,y^t\right)}{D_{VRS}^t\left(\chi^t,y^t\right)} \tag{6}$$

where, CRS represents reference technology with a constant returns to scale assumption and VRS a variable returns to scale assumption. However, Fare *et al.* (1997b) were the first to criticise the Ray & Desli (1997) model as to the fact that it cannot measure scale efficiency change (SΔ) since each component uses only single period technology (Lovell, 2001). Therefore, the equation has been extended to incorporate time and the scale efficiency change factor.

$$S\Delta\left(x^t,y^t,x^{t+1},y^{t+1}\right)=$$

$$\left[\frac{SE^t\left(x^{t+1},y^{t+1}\right)/D_0^t\left(x^{t+1},y^{t+1}\right)}{SE^t\left(x^t,y^t\right)/D_0^t\left(x^t,y^t\right)}*\frac{SE^{t+1}\left(x^{t+1},y^{t+1}\right)/D_0^{t+1}\left(x^{t+1},y^{t+1}\right)}{SE^{t+1}\left(x^t,y^t\right)/D_0^{t+1}\left(x^t,y^t\right)}\right]^{\frac{1}{2}} \tag{7}$$

Then the Malmquist productivity index ($M_o$) can be decomposed into technical efficiency change and technological change written in the first paranthesis and scale efficiency change in the latter as:

$$M_0=\left[\frac{D_{VRS}^t\left(\chi^{t+1},y^{t+1}\right)}{D_{VRS}^t\left(\chi^t,y^t\right)}\times\frac{D_{VRS}^{t+1}\left(\chi^{t+1},y^{t+1}\right)}{D_{VRS}^{t+1}\left(\chi^t,y^t\right)}\right]^{\frac{1}{2}}\times\left[\frac{SE^t\left(\chi^{t+1},y^{t+1}\right)}{SE^t\left(\chi^t,y^t\right)}\times\frac{SE^{t+1}\left(\chi^{t+1},y^{t+1}\right)}{SE^{t+1}\left(\chi^t,y^t\right)}\right]^{\frac{1}{2}} \tag{8}$$

where $D_0$ is the distance function with input vectors of $X^t$ and $X^{t+1}$ for periods t and t+1 and output vectors of $Y^t$ and $Y^{t+1}$ for periods t and t+1, respectively. $D_{VRS}^t$ and $D_{VRS}^{t+1}$ represent technologies with variable returns to scale at periods t and t+1 and $SE^t$ and $SE^{t+1}$ are variables identifying scale efficiency change at periods t and t+1. An improvement in TC is considered as a shift in the frontier. Also, scale efficiency change (SEC) component has been subject to a number of criticisms (see Casu *et al.*, 2004), mainly in terms of the role of constant returns vs. variable returns to scale frontiers. However, there seems to be consensus that the Malmquist index is correctly measured by the constant returns to scale distance function, even when technology exhibits variable returns to scale.

## 6. Empirical findings

Throughout this section, we present input oriented Malmquist Index findings obtained by the DEAP software introduced by Coelli (1996), using a panel of Turkish commercial banks with a total of 272 observations over the period 1992-2008. As aforementioned, TFPC scores are decomposed into TC, TEC, PEC and SEC components. At this stage, it may be questionable to assume that all banks are coming from the same legal and business environment and pool domestic and foreign invested banks together. As a robustness check, we performed parametric (ANOVA) and non-parametric (Wilcoxon and Kruskal-Wallis) tests to check the null hypothesis that all banks come from the same population following Isik & Hasan (2002), and Havrylchyk (2006). Results of the tests are presented in Table 4. Consistent with Sathye (2001), and Isik & Hassan (2002, 2003) tests could not reject the null hypothesis at 1% significance level, thus foreign invetsed and domestic commercial banks are from the same sample and it would be appropriate to pull all banks together using a common frontier.

| Test statistics[a] | ANOVA F (prob>F) | Wilcoxon z (prob>z) | Krusal-Wallis $\chi^2$ (prob> $\chi^2$) |
|---|---|---|---|
| **Panel A: 1992** | | | |
| TEC | 0.664 (0.112) | 0.536 (0.212) | 0.158 (0.00) |
| TC | 0.749 (0.230) | 0.344 (0.032) | 0.208 (0.00) |
| PEC | 0.663 (0.162) | 0.456 (0.788) | 0.185 (0.001) |
| SEC | 0.146 (0.01) | 0.553 (0.766) | 0.308 (0.706) |
| TFPC | 0.205 (0.233) | 0.422 (0.245) | 0.882 (0.003) |
| **Panel B: 1997** | | | |
| TEC | 0.237 (0.118) | 0.334 (0.788) | 0.118 (0.002) |
| TC | 0.201 (0.05) | 0.138 (0.03) | 0.127 (0.005) |
| PEC | 0.234 (0.046) | 0.179 (0.233) | 0.873 (0.110) |
| SEC | 0.115 (0.344) | 0.788 (0.566) | 0.432 (0.215) |
| TFPC | 0.075 (0.803) | 0.236 (0.455) | 0.576 (0.012) |
| **Panel C: 2003** | | | |
| TEC | 0.654 (0.01) | 0.765 (0.212) | 0.321 (0.056) |
| TC | 0.655 (0.01) | 0.466 (0.312) | 0.182 (0.213) |
| PEC | 1.115 (0.236) | 0.895 (0.023) | 0.193 (0.165) |
| SEC | 2.172 (0.119) | 0.896 (0.024) | 0.466 (0.344) |
| TFPC | 0.931 (0.346) | 0.269 (0.0342) | 0.877 (0.313) |

| Test statistics[a] | ANOVA F (prob>F) | Wilcoxon z (prob>z) | Krusal-Wallis $\chi^2$ (prob> $\chi^2$) |
|---|---|---|---|
| **Panel D: 2008** | | | |
| TEC | 0.030 (0.788) | 0.0988 (0.431) | 0.932 (0.122) |
| TC | 0.567 (0.345) | 0.677 (0.002) | 0.244 (0.236) |
| PEC | 0.054 (0.536) | 0.0478 (0.201) | 0.788 (0.366) |
| SEC | 2.33 (0.113) | 0.780 (0.023) | 0.780 (0.234) |
| TFPC | 0.476 (0.532) | 0.762 (0.011) | 0.566 (0.452) |

[a] Tets methodology follows Elyasiani & Mehdian (1992).
All tests applied with ownership as grouping variable. The ANOVA is a parametric test that test the null hypothesis that foreign invested and domestic banks have the same mean; Wilcoxon Rank-Sum, Kruskal-Wallis are non-parametric tests that test the shift in the location of the distribution.
The numbers in paranthesis are the p-values associated with the relevant test.
*Notes*: TEC: technical efficiency changes; TC: technological change; PEC: pure efficiency change; SEC: scale efficiency change; TFPC: total factor productivity change.

Table 4. Summary of parametric and non-parametric tests

## 6.1 DEA-Malmquist findings: Country level analysis

Table 5 presents the Malmquist TFPC estimates applying the intermediation approach by Sealey & Lindley (1977). In the last column, if the value is greater than one it indicates productivity growth, while a value less than one indicates a TFP decline over the relevant period. We report TFPC decomposed into TEC, TC, PEC and SEC as in Casu *et al.* (2004), Isik & Hassan (2003), Delis *et al.* (2008), and Dogan & Fausten (2003). This helps to isolate contributions of each component on TFPC. The annual results are geometric means, and indices of change are calculated relative to the previous year using successive reference technologies. If there is increasing returns to scale (IRS) where SEC >1 than it is optimal to expand the scale of production in order to increase productivity, whilst decrease the production level if there is DRS, i.e. SEC<1 (Isik & Hassan, 2003).

The results suggests that the average bank experienced a productivity growth of 0.6 per cent, comprising an average technical efficiency increase of 2.4 per cent, a slight average technological decline of 1.8 per cent and an average scale efficiency progress of 1.9 per cent over the 1992-2008 period. TFPC vary across the sample period with a negative change of 40.9 per cent from 1992-93 to 1993–94 attributable to the economic contraction due to the 1994 crisis, which affected the Turkish banking sector badly. In contrast, TEC experienced only 27.7 per cent decline over the same period. Followingly, an immediate recovery of 30.3 per cent is observed in TFPC during 1994–95, while the impact of the 1994 crisis affected TEC by a negative change of 16.1 per cent over the same period. We monitored frequent fluctuations in the TFPC scores of the banks under analysis with the most devastating productivity recessions during 1993–94, 1996-97 and 2000–01.

| Years | TEC | TC | PEC | SEC | TFPC |
|-------|-----|-----|-----|-----|------|
| 1992–1993 | 1.339 | 0.826 | 1.027 | 1.304 | 1.106 |
| 1993–1994 | 1.062 | 0.656 | 0.962 | 1.105 | 0.697 |
| 1994–1995 | 0.901 | 1.109 | 0.947 | 0.952 | 1.000 |
| 1995–1996 | 1.003 | 1.127 | 1.001 | 1.002 | 1.130 |
| 1996–1997 | 1.030 | 0.666 | 1.041 | 0.989 | 0.686 |
| 1997–1998 | 1.110 | 0.656 | 1.117 | 0.994 | 0.728 |
| 1998–1999 | 0.875 | 0.912 | 0.946 | 0.926 | 0.798 |
| 1999–2000 | 1.092 | 1.245 | 1.069 | 1.022 | 1.359 |
| 2000–2001 | 1.053 | 0.541 | 1.012 | 1.041 | 0.569 |
| 2001–2002 | 0.885 | 1.380 | 0.863 | 1.026 | 1.221 |
| 2002–2003 | 0.965 | 1.357 | 1.011 | 0.954 | 1.309 |
| 2003–2004 | 0.917 | 1.305 | 0.968 | 0.947 | 1.196 |
| 2004–2005 | 0.965 | 1.888 | 0.984 | 0.981 | 1.822 |
| 2005–2006 | 1.053 | 0.990 | 1.113 | 0.947 | 1.043 |
| 2007–2007 | 1.239 | 0.814 | 1.045 | 1.186 | 1.008 |
| 2007–2008 | 1.001 | 1.113 | 1.000 | 1.001 | 1.114 |
| Mean | 1.024 | 0.982 | 1.005 | 1.019 | 1.006 |

*Notes*: 1 = 100 per cent; TEC: technical efficiency change; TC: technological change; PEC: pure efficiency change; SEC: scale efficiency change; TFPC: total factor productivity change.

Table 5. DEA-Malmquist components, 1992–2008 (1=100 per cent)

Starting with 1996, banks in the sample experienced a negative TFPC until 1999 followed by a one year productivity growth in the 1999-2000 period. We suggest the negative trend over 1996-99 period as a contagion effect of Russian crisis in 1998 as it is one of the largest trade partners of Turkey. Attributable to the twin crises of 2000-01, TFPC was negative which suggests the fact that banks could not cope with the changing economic and technological environment that arose from the twin crises and know-how transferred into the system by foreign investors; and since then TFPC between any successive years is consistently positive.

The significant positive TFPC starting in 2001-02 period may be correlated with the signing of the tough IMF recovery program in 2002, and entry of foreign banks into the system, for instance sale of Demirbank to HSBC, and acquisition of Kocbank with UniCredito during 2001 and 2002, respectively. In connection with Eller *et al.* (2005) this may be the result of better management of financial resources to high return projects, technical progress and better risk diversification of foreign-invested banks. For instance, during 2003-04, the productivity growth (19.6 per cent) was primarily the result of technological progress, TC (30.5 per cent) that was partly offset by technical inefficiency (8.3 per cent). Taken together, these results suggest that unproductive banks tend to catch up with productive banks largely in terms TC as banks experienced recessions in their TEC estimates. Over the analysis period, we suggest that the predominant source of TFPC is TC confirmed by the significantly positive Pearson and Kendall's correlation coefficients introduced in Table 6.

|  | TEC[b] | TC[b] | PEC[b] | SEC[b] | TFPC[b] |
|---|---|---|---|---|---|
| **TEC** |  |  |  |  |  |
| p | 1.0000*** |  |  |  |  |
| k | 1.0000*** |  |  |  |  |
| **TC** |  |  |  |  |  |
| p | 0.2617*** | 1.0000*** |  |  |  |
| k | -0.1785** | 1.0000*** |  |  |  |
| **PEC** |  |  |  |  |  |
| p | 0.7114*** | -0.0561* | 1.0000*** |  |  |
| k | 0.5530*** | -0.0383* | 1.0000*** |  |  |
| **SEC** |  |  |  |  |  |
| p | 0.5134*** | -0.2567** | -0.080*** | 1.0000*** |  |
| k | 0.3793*** | -0.1720** | -0.0538** | 1.0000*** |  |
| **TFPC** |  |  |  |  |  |
| p | 0.4253*** | 0.6962*** | 0.4362*** | 0.0561*** | 1.0000*** |
| k | 0.2922** | 0.5292*** | 0.2891*** | 0.0367*** | 1.0000*** |

[a] Parametric ordinary Pearson (p) correlation coefficients – first rows of each cell. Kendall's (k) correlation coefficients–second rows of each cell.
[b] TEC= Technical efficiency change, TC= Technological change, PEC= Pure efficiency change, SEC= Scale efficiency change, TFPC= Total factor productivity change,
*, ** and *** indicate significance levels of 10%, 5% and 1%, respectively.

Table 6. Pearson and Kendall's rank order correlation coefficients

The null hypothesis is that the correlation coefficient between two variables is zero. In all cases, the Pearson (p) coefficient results confirm all the relationships found with the Kendall's (k) in the direction (positive or negatave) and significance. All of the DEA-Malmquist decomposed components, namely, TEC, TC, PEC and SEC, are positively correlated with TFPC ($p_{TFPC-TEC}$=0.4253, $p_{TFPC-TC}$=0.6962, $p_{TFPC-PEC}$=0.4362, $p_{TFPC-SEC}$=0.0561, respectively). TC and PEC are highly positively and statistically significantly associated with TFPC indicating the dominant effect of technological change and managerial efficiency change on the overall productivity scores of banks. TEC is more related to PEC and SEC than to TC ($p_{TEC-PEC}$=0.7114, $p_{TEC-SEC}$=0.5134, $p_{TEC-TC}$=0.2617, respectively), confirming the dominant effect of managerial efficiency and scale efficiency in determining the technical efficiency of the Turkish commercial banks. This finding also confirms a similar statement by Isik & Hassan (2002).

It can be stated that, high-tech investments played a crucial role in the positive TC of banks and also on productivity growth over the analysis period. We suggest this as the result of rapid spread of Automatic Telling Machines (ATMs), Point on Sale (POS) terminals and increased number of bank cards issued as a reflection of the widespread acceptance gained by these products. High-tech investments brought into the system by foreign-invested banks is suggested to play a crucial role on positive TC shifting the frontier upwards.

Moreover, significantly positive and high managerial efficiency (PEC) since 2005-06 period can be attributable to the argument that foreign invested banks may increase the quality of human capital in a banking system, either by importing high skilled bank managers to work in their branches or by training the local employees (Lensink & Hermes, 2004). The DEA-Malmquist results also indicated a trade-off between managerial efficiency and technical progress over the period of analysis, which could be explained by the fact that technology transfer through FDI takes time to materialise as resources have to be devoted to learning during which time banks seek ways of attaining gains in terms of managerial or scale efficiency.

## 6.2 DEA-Malmquist findings: Bank level analysis

Our findings in Table 7 suggest that over the 1992-2008 period, foreign-invested banks achieved higher average TFP growth than average of 0.6 per cent for the whole commercial banks in the sample. The highest average achievements in TFPC are experienced by foreign-invested Akbank (investment by Citibank), YKB-UniCredito (earlier Kocbank), Sekerbank (investment by BTA) and Garanti bank (investment by GECC), which are well above the average. Koçbank presented the highest average productivity growth (39 per cent) - prior to the acqusition by YKB. Looking at the figures, we observe that Kocbank experienced the highest average TFPC and TC over the analysis period.This can be attributable to its high level of performance from 2000-01 to 2002-03. This period coincides with the point in time of its merger with the Italian bank, UniCredito, in 2002.

Annual TFPC scores of Dışbank have been the most volatile of all. It has presented a pattern of great sensitivity to macroeconomic and policy changes in the country, which can be a sign that it was not as powerful a bank as were its counterparts under focus. It's most recent recovery (31 per cent) was identified after its sale to one of the most reliable banks of Europe, namely Fortis Bank of Belgium, in 2005. Contrary to the productivity growth experienced during various periods, Kocbank and Disbank experienced productivity decline during the financial crisis periods in the country. We attribute high productivity scores of foreign-invested banks in Turkey to their high level of TC compared to domestic banks. This may suggest that foreign banks which have invested in local Turkish banks have succeeded in utilizing their superior technology and expertise resulting in productivity growth higher than the market average and domestic banks.

The same is applicable for TEC estimates but this time adding Finansbank to the list. In general, TEC results indicate that foreign-invested banks are relatively more efficient (higher postive TEC) than their domestic counterparts consistent with the findings of Hasan & Hunter (1996), Mahajan et al. (1996), DeYoung & Nolle (1996) and Chang et al. (1998).

| Bank | TEC | TC | PEC | SEC | TFPC |
|------|-----|-----|-----|-----|------|
| Ziraat Bankası | 0.982 | 0.992 | 0.947 | 1.037 | 0.974 |
| Halk Bankası | 1.046 | 0.985 | 1.002 | 1.043 | 1.031 |
| Vakıflar Bankası | 1.090 | 1.029 | 1.044 | 1.045 | 1.122 |
| Alternatif Bank | 0.985 | 0.951 | 0.985 | 1.000 | 0.937 |
| Anadolu Bank | 0.984 | 0.811 | 0.984 | 1.000 | 0.789 |
| Şekerbank-BTA | 1.062 | 1.005 | 1.040 | 1.021 | 1.067 |
| Tekstil Bank | 1.000 | 0.991 | 0.993 | 1.007 | 0.991 |
| TEB-BNP Paribas | 1.008 | 0.989 | 1.003 | 1.005 | 0.997 |
| Garanti Bankası-GECC | 1.041 | 1.020 | 1.022 | 1.019 | 1.062 |
| Işbankası | 1.040 | 1.015 | 1.004 | 1.036 | 1.056 |
| YKB-UniCredito | 1.036 | 1.002 | 0.997 | 1.039 | 1.038 |
| Akbank-Citibank | 1.064 | 1.001 | 1.038 | 1.026 | 1.065 |
| Koçbank-Unicredit/YKB | 1.031 | 1.347 | 1.030 | 1.002 | 1.390 |
| Dışbank-Fortis | 1.000 | 0.951 | 1.000 | 1.000 | 0.951 |
| Demirbank-HSBC | 1.012 | 0.981 | 0.993 | 1.019 | 0.993 |
| Denizbank-Dexia Bank | 1.000 | 0.777 | 1.000 | 1.000 | 0.777 |
| Finansbank-NBG | 1.031 | 0.952 | 1.000 | 1.031 | 0.981 |
| Mean | 1.024 | 0.982 | 1.005 | 1.019 | 1.006 |

Notes: TEC: technical efficiency changes; TC: technological change; PEC: pure efficiency change; SEC: scale efficiency change; TFPC: total factor productivity change.

Table 7. DEA-Malmquist decomposed components of individual banks, 1992–2008

Another comment was suggested by Havrylchyk (2006), who emphasize that foreign banks acquire more efficient banks in a banking industry, whilst fail to enhance their efficiency further.

In terms of public ownership, a noteworthy aspect of public banks is related to credit. As a result of the political pressure in Turkey, public banks - Ziraat Bankası, Halk Bankası and Vakıflar Bankası - issued loans more easily than their private counterparts. This helped to increase their level of output making them appear efficient and productive. However, this may not be the case on a risk-adjusted basis, as public banks carry a large number of non-performing loans. If adjustments to their outputs were made to reflect loan losses, public banks might in fact be found to be less efficient than private banks under examination. Nevertheless, lack of detailed data on bad loans of public banks prevents an in-depth examination of this hypothesis.

Table 8 presents the pre and post FDI technological change scores of foreign invested banks. Estimates highlighted with bold represents the period of post-FDI. It is observed that except from Denizbank, all foreign invested private banks in Turkey have experienced productivity growth during the pre-FDI periods. This supports the argument that efficiency or productivity is a pre condition to FDI and foreign investors target good performing banks to invest.

| | 1992-2000 | 2000-04 | 2004-05 | 2005-06 | 2006-07 | 2007-08 |
|---|---|---|---|---|---|---|
| Sekerbank | 0.933 | 1.254 | 1.274 | 1.07 | **0.823** | **0.999** |
| TEB [*] | 0.905 | 1.245 | 1.454 | **1.008** | **0.766** | **1.017** |
| Garanti Bankası | 0.994 | 1.190 | 1.449 | **1.028** | **0.773** | **1.102** |
| YKB | 0.945 | 1.196 | 1.333 | **1.031** | **0.783** | **1.07** |
| Akbank | 0.931 | 1.369 | 1.314 | 1.06 | **0.839** | **1.065** |
| Kocbank | 0.981 | **1.182** | **1.262** | - | - | - |
| Dışbank | 0.991 | 0.999 | 1.243 | **1.104** | **0.89** | **1.002** |
| Demirbank | 0.945 | **1.099** | **1.237** | **1.101** | **0.904** | **0.999** |
| Denizbank | 0.866 | 1.033 | 1.3633 | **0.442** | **0.773** | **1.045** |
| Finansbank | 0.932 | 1.095 | 1.228 | **1.081** | **0.826** | **1.003** |

Numbers in bold represents the post-FDI periods.

Table 8. Pre and post-FDI technological change of foreign invested banks

# 7. Conclusions

Employing the input oriented DEA-Malmquist model, we estimate the total factor productivity change scores of 17 Turkish commercial banks over the period 1992–2008 period to study the impact of foreign ownership on total factor productivity change and technological change scores of Turkish banking sector. Over the years under study, empirical results indicate that the productivity scores of the industry consistently fell over the crises years. The mean productivity estimates that we find for foreign invested banks in general are higher than those of domestic banks.

We also decompose the productivity change scores (TFPC) into its technical efficiency change (TEC), technological change (TC), pure efficiency change (PEC) and scale efficiency change (SEC) components. Overall, the decomposed figures indicate that the most significant factor on the TFPC scores is the TC component. The major investments in high tech bank operations shifted the frontier upward, in particular after the entry of foreign banks in the sector. This is due to the fact that foreign involvement created pressure on the Turkish banks as a result of increased competition and forced them to diminish their costs. In particular, Demirbank, which was sold to HSBC in 2001, experienced massive productivity growth in the post-FDI period. Despite its beneficial effects and increased trend since 2005, recorded FDI inflows to Turkey have been exceptionally low compared with those of the central and east European countries. The main FDI challenges facing Turkey are determining why FDI inflows have remained so low and how Turkey can increase the inflows to desirable levels.

In particular, stability and trust made Turkey an attracting market for foreign investors giving rise to further economic growth during the recent years. In this respect, Turkey with an increasing population, increasing per capita income and having an advantage of geographical location connecting Europe and Middle East is expected to be a further magnet for foreign investors. Moreover, Turkey is in the EU accession process and Turkish banks are in general smaller in size compared to European counterparts, making them easy to be taken over. Also, the flexibility in the Turkish Banking Law treats both domestic and foreign banks the same and does not put limitations on the share of foreign ownership. All of these important issues can be suggested as the key factors boosting up FDI in the Turkish banking sector.

From the economic perspective, it can be suggested that Turkey's accession to the EU should be related to its size, per capita income, and dependence on agriculture. For the EU, these three factors combine to create a huge immigration potential if migration is let free. Moreover, these factors indicate that Turkey may become the largest recipient of transfers from the EU budget due to its large population, at least under the present rules and policies. As a result of these factors, Turkey will face further challenges from the current EU member countries. Given the fact that its size is almost as big as the total of new member countries (EU-12), Turkey as a single country will have strong voting rights in the European Council. This is a big question mark in the EU's mind whether to approve Turkey's accession or not.

## 8. References

Aitken, B. J. & Harrison, A. E. (1999). Do Domestic Firms Benefit from Direct Foreign Investment? Evidence from Venezuela. *American Economic Review*, American Economic Association, Vol. 89, No. 3, pp. 605-618.

Altunbas, Y., Molyneux, P., & Murphy, N. (1994c). Privatization, efficiency and public ownership in Turkey – An analysis of the banking industry 1991-1993. Unpublished working paper, Institute of European Finance.

Aysan, A.F. & Ceyhan, S.P. (2008). What determines the banking sector performance in globalized financial markets? The case of Turkey, *Physica A*, No. 387, pp. 1593-1602.

BAT: Quarterly Statistics by Banks, Banks Association of Turkey, various years http://www.tbb.org.tr

Berg, S., Forsund, F. & Jansen, E. (1992). Malmquist indices of productivity growth during the deregulation of Norwegian banking, 1980-1989. *Scandinavian Journal of Economics*, Vol. 94, pp. 211-228.

Berger, A. & Humphrey, D. (1997). Efficiency of Financial Institutions: International Survey and Directions for Future Research. *European Journal of Operational Research*, Vol. 98, No.2, pp. 175-212.

Berger, A.N. & Mester, L.J. (2003). Explaining the dramatic changes in performance of US banks: technological change, deregulation, and dynamic changes in competition. *Journal of Financial Intermediation*, Vol. 12, pp. 57-95.

Berger, A.N., & Mester, L.J. (1997). Inside the black box: What explains differences in the efficiencies of financial institutions?. *Journal of Banking & Finance*, Vol. 21, pp. 895–947.

Berger, A.N., Deyoung, R., Genay, H. & Udell, G.F. (2000). Globalization of Financial Institutions: Evidence from Cross-Border Banking Performance. *Brookings-Wharton Papers on Financial Services*, pp. 23–120

Bonin, J.P., Hasan, I., & Wachtel, P. (2005). Bank performance, efficiency and ownership in transition countries. *Journal of Banking & Finance*, Vol. 29, pp. 31–53.

Casu, B., Girardone, C. & Molyneux, P. (2004). Productivity Change in European Banking: A Comparison of Parametric and Non-parametric Approaches. *Journal of Banking & Finance*, Vol. 28, No. 10, pp. 2521-2540

Cavallo, L. & Rossi, S.P.S. (2002). Do environmental variables affect the performance and technical efficiency of European banking systems? A parametric analysis using the stochastic frontier approach, *European Journal of Finance*, Vol. 8, pp. 123–146

Caves, D. W. Christensen, L. R. and Diewert, W. E. (1982b). The Economic Theory of Index Numbers and the Measurement of Input, Output and Productivity. *Econometrica*, Vol. 50, pp. 1393–414.

Chaffai, M. E., Dietch, M., & Lozano-Vivas, A. (2001). Technological and Environmental Differences in the European Banking Industries. *Journal of Financial Services Research*, Vol. 19, No. 2/3, pp. 147–62.

Chang, C., Hasan, I., & Hunter, W. (1998). Efficiency of multinational banks: An empirical investigation. *Applied Financial Economics*, Vol. 8, pp. 689–696.

Claessens, S., Demirguc-Kunt, A. & Huizinga, H. (2001). How does foreign entry affect domestic banking markets?. *Journal of Banking and Finance*, Vol. 25, pp. 891-911.

Coelli, T. (1996), A guide to DEAP version 2.1: a Data Envelopment Analysis (Computer programme). *CEPA Working Paper 96/08*, Department of Econometrics, University of New England, Armidale, Australia.

Delis, M.D., Staikouras, C.K. & Varlagas, P. (2008). On the Measurement of Market Power in the Banking Industry. *Journal of Business Finance and Accounting*, forthcoming.

Demirguc-Kunt, A. & Huizinga, H. (2000). Determinants of commercial bank interest margins and profitability: Some international evidence. *World Bank Economic Review*, No. 13, pp. 379–408.

Denizer, C.A., Dinc, M. & Tarimcilar, M. (2000). Measuring banking efficiency in the pre- and postliberalisation environment: Evidence from the Turkish banking system. *World Bank Working Paper*, No. 2476.

DeYoung, R. & Nolle, D.E. (1996). Foreign-owned banks in the US: Earning market share or buying it?. *Journal of Money, Credit and Banking*, Vol. 28, pp. 622-636.

Dietsch, M. & Lozano-Vivas, A. (2000). How the environment determines banking efficiency: a comparison between French and Spanish industries". *Journal of Banking & Finance*, Vol. 24, No. 6, pp. 985–1004.

Dogan, E. & Fausten, D. (2003). Productivity and technical change in Malaysian banking: 1989–1998, *Asia-Pacific Financial Markets*, Vol. 10, pp. 205–237

Eller, M., Haiss, P. & Steiner, K. (2005). Foreign direct investment in the Financial Sector: The Engine of Growth, *EI Working Paper, No. 69*

Elyasiani, E. & Mehdian, S. M. (1992). Productive Efficiency Performance of Minority and Nonminority-Owned Banks: A Nonparametric Approach. *Journal of Banking and Finance*, Vol. 16, pp. 933–948.

Färe, R, Grosskopf, S., Norris, M. & Zhang, Z. (1994). Productivity Growth, Technical Progress, and Efficiency Change in Industrialized Countries. *American Economic Review*, Vol. 84, pp. 66–83.

Färe, R., Grifell-Tatjé, E., Grosskopf, S. & Lovell, C. A. K. (1997b). Biased Technical Change and the Malmquist Productivity Index. *Scandinavian Journal of Economics*, Vol. 99, No. 1, pp. 119–27.

Farrell, M. J. (1957). The Measurement of Productive Efficiency. *Journal of the Royal Statistical Society*, Vol. 120, pp. 253–81.

Fries, S. & Taci, A. (2002). Banking Reform and Development in Transition Economies. *EBRD Working Paper*, No. 71.

Global Competitiveness Report 2007-2008, World Economic Form: October 31 2007.

Görg, H. & Greenaway, D. (2004). Much Ado about Nothing? Do Domestic Firms Really Benefit from Foreign Direct Investment?. *World Bank Research Observer*, Oxford University Press, Vol. 19, No. 2, pp. 171-197.

Grifell-Tatje, E. & Lovell, C.A.K. (1997). The sources of productivity change in

Hancock, D., Humphrey, D.B. & Wilcox, J.A. (1999). Cost reductions in electronic payments: The roles of consolidation, economies of scale and technical change, *Journal of Banking and Finance*, Vol. 23, pp. 391-421

Hasan, I., & Hunter, W.C. (1996). Efficiency of Japanese multinational banks in the United States. In: Chen, Andrew H. (Ed.), Research in Finance, Vol. 14. JAI Press, Greenwich, CT and London, pp. 157– 173.

Havrylchyk, O. (2006). Efficiency of the Polish banking industry: foreign versus domestic banks. *Journal of Banking and Finance*, Vol. 30, pp. 1975-1996

Hymer, S. (1976). The International Operations of National Firms: A Study of Direct Foreign Investment. MIT Press, Cambridge, Massachussets.

Isik, I. & Hasan, M.K. (2002). Technical, Scale and Allocative Efficiencies of Turkish Banking Industry. *Journal of Banking and Finance*, Vol. 26, No. 4, pp. 719-766.

Isik, I. (2007). Bank ownership and productivity developments: evidence from Turkey. *Studies in Economics and Finance*, Vol. 24, pp. 115-139.

Isik, I., & Hassan, M. K. (2003). Financial disruption and bank productivity: The 1994 experience of Turkish banks. *The Quarterly Review of Economics and Finance*, Vol. 43, pp. 291-320.

Kraft, E. & Tirtiroglu, D. (1998). Bank efficiency in Croatia: A stochastic-frontier analysis. *Journal of Comparative Economics*, Vol. 26, pp. 282–300.

Lensink, R., Meesters, A. & Naaborg, I. (2008). Bank efficiency and foreign ownership: do good institutions matters?. *Journal of Banking and Finance*, Vol. 5, pp. 834-44.

Liu, Z. (2008). Foreign direct investment and technology spillovers: theory and evidence. *Journal of Development Economics*, Vol. 85, pp. 176-93.

Lovell, K. (2001). Future Research Opportunities in Efficiency and Productivity Analysis. *Efficiency Series Paper*, No. 1, University of Oviedo Department of Economics.

Lozano-Vivas, A., Pastor, J. A. & Pastor, J. M. (2002). An Efficiency Comparison of European Banking Systems Operating under Different Environmental Conditions. *Journal of Productivity Analysis*, Vol. 18, pp. 59–77.

Mahajan, A., Rangan, N. & Zardkoohi, A. (1996), Cost Structure in Multinational and Domestic Banking. *Journal of Banking and Finance*, Vol. 20, pp. 283-306.

Malmquist, S. (1953). Index Numbers and Indifference Surfaces. *Trabajos de Estatistica*, Vol. 4, pp. 209–42.

Mercan, M., Reisman, A., Yolalan, R., & Emel, A.B. (2003). The effect of scale and mode of ownership on the financial performance of the Turkish banking sector: Results of a DEA based analysis. *Socio-Economic Planning Sciences*, Vol. 37, No. 3, pp. 185–202.

Mian, A., (2006). The limits of foreign lending in poor countries. *The Journal of Finance*, Vol. 61, pp. 1465–1505.

Moran, T.H., Graham, E.M. & Blomstrom, M. (Eds) (2005). Does Foreign Direct Investment Promote Development?. Peterson Institute of International Economics,Washington, DC, May, pp. 440.

Mukerjee, K., Ray, S.C. & Miller, S.M. (2001). Productivity growth in large US commercial banks: the initial post-deregulation experience. *Journal of Banking and Finance*, Vol. 25, pp. 913-39.

Oral, M., & Yolalan, R. (1990). An empirical study on measuring operating efficiency and profitability of bank branches. *European Journal of Operational Research*, Vol. 46, pp. 282–294.

Pasiouras, F. (2008). International evidence on the impact of regulations and supervision on banks' technical efficiency: an application of two-stage data envelopment analysis. *Review of Quantitative Finance and Accounting*, Vol. 30, pp. 187-223.

Pasiouras, F., Tanna, S. & Zopounidis, C. (2007). Regulations, supervision, and banks' cost and profit efficiency around the world: a stochastic frontier approach. working paper series, 2007.05, University of Bath School of Management, Bath.

Ray, S. & Desli, E. (1997). Productivity Growth, Technical Progress, and Efficiency Change in Industrialized Countries: Comment. *American Economic Review*, Vol. 87, No. 5, pp. 1033–9.

Rezvanian, R., & Mehdian, S. (2002). An examination of cost structure and production performance of commercial banks in Singapore. *Journal of Banking and Finance*, Vol. 26, pp. 79–98.

Sathye, M. (2001). X-efficiency of Australian banking: An empirical investigation. *Journal of Banking and Finance*, Vol. 25,pp. 613–630.

Sealey Jr., C.W. & Lindley, J.T. (1977). Inputs, outputs, and a theory of production and cost at depository financial institutions. *Journal of Finance*, Vol. 32, pp. 1251–1266.

Spanish banking. *European Journal of Operational Research*, Vol. 5, pp. 257-264.

Stein, J. (2002). Information Production and Capital Allocation: Decentralized versus Hierarchical Firms. *Journal of Finance*, Vol. LVII, No.5, October 2002.

Steinherr, A., Tukel, A. & Ucer, M. (2004). The Turkish Banking Sector: Challenges and Outlook in Transition to EU Membership. *Bruges European Economic Policy Briefings*, No.9.

Tirtiroglu, D., Daniels, K.N. & Tirtigoglu, E. (2005). Deregulation, Intensity of Competition, Industry Evolution, and the Productivity Growth of U.S. Commercial Banks. *Journal of Money, Credit and Banking*, Vol. 37, No. 2, pp. 339-360.

Worthington, A.C. (1999). Malmquist indices of productivity change in Australian financial services. *Journal of International Financial Markets, Institutions and Money*, Vol. 9, pp. 303-320.

Yildirim, C. (2002). Evolution of banking efficiency within an unstable macroeconomic environment: The case of Turkish commercial banks. *Applied Economics*, Vol. 34, pp. 2289-2301.

Yolalan, R. (1996). Turk Bankacilik Sektoru icin Goreli Mali Performans Olcumu. *Turkiye Bankalar Birligi Bankacilar Dergisi*, Vol. 19, pp. 35-40.,

Zaim, O. (1995). The effect of financial liberalisation on the efficiency of Turkish commercial banks. Applied Financial Economics, Vol. 5, pp. 257-264

# 11

# Knowledge Frontiers for Sustainable Growth and Development in Zimbabwe

Gabriel Kabanda

*Zimbabwe Open University, Harare*
*Zimbabwe*

## 1. Introduction

Knowledge is increasingly driving growth and transforming nations and the way of life. The essence of sustainable development in Africa commands a dramatic reduction of poverty and hunger and improved development prospects for future generations (Hamel, 2004). Achievement of meaningful sustainable development requires considerable progress on these fronts. Knowledge-driven sustainable development requires relevant and efficient development knowledge. *"Sustainable development"* is development that meets the needs of the present without compromising the ability of future generations to meet their own needs[1].

Sustainable development has remained elusive for many African countries. Africa's efforts to achieve sustainable development have been hindered by conflicts, insufficient investment, limited market access opportunities and supply side constraints, unsustainable debt burdens, historically declining levels of official development assistance and the impact of HIV/AIDS[2]. Zimbabwe is faced with the following three major sustainable development problems that require knowledge frontiers as a solution:

1. Attainment of the Millenium Development Goals (MDGs)
2. Poverty reduction
3. Social problems value chain

Knowledge has become the new currency in the modern age and in spearheading Zimbabwe to become a knowledge-based economy. On the contrary, knowledge deficiencies and shortcomings affect various forms of sustainable development (social, cultural, economic, technological, political, and anthropological). Zimbabwe, like other African countries, should establish African Knowledge Systems (AKSs) from which knowledge policies may precipitate. Knowledge is the centrality of sustainable growth and development in many ways, which is the main object of this paper. To some extent, knowledge for sustainable development is something knowable, modelable, reformable and manageable through effective knowledge policies (Hamel, 2004). However, sufficient

---

[1] World Commission on Environment and Development (WCED). *Our Common future.* Oxford. Oxford University Press, 1987 p.43.
[2] http://www.un.org/esa/sustdev/documents/WSSD_POI_PD/English/POIChapter8.htm.

understanding of knowledge in the context of an African knowledge corpus and AKSs brings some completeness in the achievement of sustainable growth and development. The changing global geography of knowledge and the character and perspectives of an AKS are uncovered in this case study for Zimbabwe with respect to:

- Knowledge for modernization and development
- Indigenous and traditional knowledge, and
- Faith-based knowledge.

The frontiers of knowledge are boundless, producing wealth for both entrepreneur and economy for *knowledge creation, commercialisation and innovation*. Recognising the critical need for knowledge as input, Zimbabwe has embarked on the transformation from an input-driven growth strategy that had served her well since political independence in 1980, so that our economy is increasingly driven by knowledge in order to achieve sustainable high growth and development. The intention to migrate from a production-based economy to a knowledge-based economy and the development of a Master Plan to chart the strategic direction towards the knowledge-based economy are now critical in Zimbabwe. The paper has a focus on the exploration of frontiers in order to encourage original innovation, integrated innovation and re-innovation based on the absorption and digestion of existing knowledge for sustainable growth and development. The New Partnership for Africa's Development (NEPAD) is a commitment by African leaders to the people of Africa and which provides a framework for sustainable development on the continent to be shared by all Africa's people. The four pillars of economic growth are:

- An educated and skilled population to create, share, and use knowledge well.
- A dynamic information infrastructure to facilitate the effective communication, dissemination, and processing of information.
- Anefficient innovation system comprising academia, firms, consultants, SMEs, etc.
- An enabling environment with supportive economic and institutional mechanisms.

## 1.1 Millenium development goals

The strategy for sustainable growth and development in Zimbabwe is to build a knowledge economy with more competitiveness and social cohesion. Knowledge is possibly the greatest producer of wealth, affluence, prosperity and 'development', and also of pollution, destruction, poverty, hunger and inequality between humans (Jamel, 2004). These are the most critical issues of sustainable development. This exciting wave of new knowledge provides knowledge opportunities for the sustainable development of the African continent, perhaps through adequate knowledge policies.

The Millenium Development Goals (MDGs) are illustrated in Figure 1 below. The ability to generate and harness sophisticated levels of knowledge is a necessary and sufficient condition to the attainment of the MDGs. Goal 1 (eradicate extreme poverty and hunger) directly affects Goals 2 to 7, and requires Goal 8 (develop a global partnership for development). According to World Development Indicators (2004), Sub-Saharan Africa (SSA) lags far behind in growth of income per capita with levels of 0.2% and 0.3% for the periods 1965-1990 and 1990-2003, respectively, as illustrated below in Figure 2. The key African problem on development is characterised by a world contribution to knowledge of only 0.03%, average tertiary enrolment ratio for school leavers of only 4% and university

Fig. 1. MDGs and Knowledge

fields of study largely dominated by social sciences/humanities with 47%, education 22%, agriculture 3%, sciences 9% and health sciences 9% (UNESCO, 2010).

The major problem of under-development characterised by the huge challenge to achieve the Millennium Development Goals (MDGs) is on knowledge empowerment mainly supported by Information and Communication Technologies (ICTs). The emergence and convergence of information and communication technologies (ICTs) has remained at the centre of global socio-economic transformations. As noted by the World Bank (2003), the effective use of technology is dependent not only on the technology but also on factors that are independent of the technology. ICTs increase productivity through:

- Better communication and networking at lower costs
- Digitalisation of production and distribution
- New trade opportunities through e-commerce
- Access to knowledge
- Increased competition

ICTs impact on all the MDGs in different ways (Kabanda, 2011). The fast track to the achievement of MDGs lies greatly in the ability to effectively manage the diffusion and adoption of ICTs for development.Debates have ensured on how information and communication technologies (ICTs) can help to alleviate poverty in low-income countries

# Sub-Saharan Africa lags far behind in growth of income per capita...

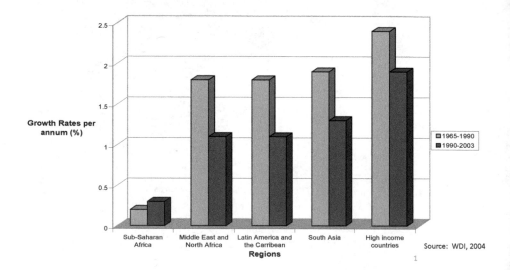

Fig. 2. Growth of income per capita in Sub-Saharan Africa

(Heeks, 1999). Advances in communication technologies have enabled many countries to improve the lives of their citizens through improved health, education and public service systems, and economies (Kekana, 2002). A knowledge economy requires:

- widespread access to communication networks;
- the existence of an educated labour-force and consumers (human capital); and
- the availability of institutions that promote knowledge creation and dissemination.

### 1.2 Poverty

Research and technology are, together with education and innovation, the components of the "Triangle of knowledge". Rich in human capital and natural resources, Zimbabwe can achieve sustainable growth and development. There is clear recognition that knowledge, not just science, technology and innovation, performs an important role in improving rates of development and growth. The solution to poverty is therefore knowledge, as illustrated on Figure 3 below.

A formalization of the impact of knowledge on growth, knowledge accumulation and application are necessary conditions that contribute to increased economic development and are at the core of a country's competitive advantage in the global economy (World Bank, 2002).

# The solution to <u>Poverty</u> is <u>Knowledge</u> for development

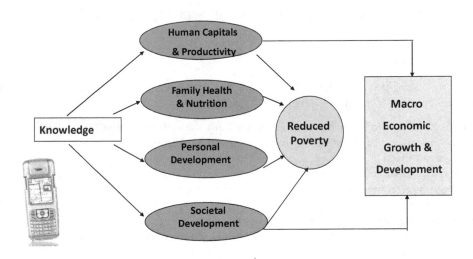

Fig. 3. Poverty and Knowledge for sustainable development

## 1.3 Social problems

The perspective on knowledge as a sustainable resource for development indicates that knowledge is a necessary condition for sustainable development in various constructed environments. Political environments require full universal access to relevant knowledge for making sound political choices and for meaningful, participative and democratic governance. Cultural environments require the full utilization of all talents and available knowledge, and a diversity of knowledge sources, including from women. Economic environments require full access to modern and efficient development knowledge. Commercially exploited knowledge is well justified in some environments (Jamel, 2004).

A social change value chain links people, communities, money and institutional capacity through knowledge frontiers, perhaps through unmet social needs. **The social** change value chain of Zimbabwe is faced with the following ethical problems (Kurasha, 2011):

1.  Domestic violence
2.  Corruption
3.  Digital illiteracy
4.  Abuse of public company and position
5.  High crime rate

6.  Child abuse and child labour
7.  Lack of institutional commitment
8.  Tribalism and other forms of segregation
9.  Sexism
10. Lack of vision
11. Political Violence
12. Pollution

The solution to all these problems of the social value chain is knowledge for the Zimbabwean societies.

## 2. The research problem

The *major sustainable developmental problems* being faced by Zimbabwe are multi-faceted and include the following symptoms (Kabanda, 2011):

1.  Many donor-driven initiatives that excluded both policy-formulation frameworks and sustainable capacity building have not brought meaningful development in these African countries.
2.  The Government policies, donor interest and community development needs are totally divergent with respect to priority areas for development.
3.  The under-development, poverty and illiterate cycles in Africa need to be broken in the long term and exploit the blessed resources available to create wealth. Extensive investment in technology and human capital development as a vehicle to exploit the vast mineral and natural resources has not been given sufficient attention.
4.  Poverty reduction requires a sustainable solution that increases production capacity at individual, institutional, community and national levels. The impact of knowledge frontiers on MDGs and generally economic growth needs a detailed assessment.

The specific objectives of the research are to:

1.  Assess the major sustainable development problems:
    a.  Attainment of the Millenium Development Goals (MDGs)
    b.  Poverty reduction
    c.  Social problems value chain
2.  Explore the emerging knowledge frontiers for sustainable growth and development in Zimbabwe.
3.  Ascertain the impact of knowledge frontiers on economic growth, innovations and education in Zimbabwe.

## 3. Literature review

In the modern world of continuous social and technological change, the demand for a workforce with capability of adjusting and attaining new skills constantly is increasing phenomenally. For example, investigative skills, thinking critically, working independently with others, and application of knowledge and skills to different situations and subject to constant change. However, knowledge can be viewed alongside a continuum: data – information – knowledge - wisdom.

## 3.1 African knowledge for sustainable development

Knowledge is extraordinarily elusive, indefinable, versatile, multipurpose, multifaceted, diverse, incommensurable, and heterogeneous (Jamel, 2004). However, knowledge is durable although its relevance or effectiveness may diminish. Development knowledge refers broadly to the totality of representative mental or abstract structures and constructions related to sustainable development. Knowledge is closely associated with beliefs and values. A good understanding of the complexity, diversity, heterogeneity, incommensurability, multidimensionality, uncertainty, comprehensibility, relativity and knowability of knowledge may facilitate the formulation of effective policies for knowledge-oriented sustainable development. Symbolical, mythological, magical, metaphorical, proverbial, and poetical knowledge pervades AKSs and is the social cement that holds them together.

## 3.2 Knowledge management

To attain business competitiveness, there is a great need for skilled knowledge workers and methods of managing the knowledge production by people, processes and business technologies. Future economic competiveness is premised on knowledge management. The function of knowledge management is to allow an organisation to leverage its information resources and knowledge assets by remembering and applying experience (Watson, 2003). Through knowledge management, organisations can become more efficient and effective as employees share knowledge and learn from one another. *Knowledge management* (KM) is "a deliberate, systematic business optimisation strategy that selects, distils, stores, organises, packages and communicates information essential to the business of a company in a manner that improves performance and corporate competitiveness" (Bergeron, 2003). However, Malhotra (1998, 2005) cites that KM "embodies organisational processes that seek synergistic combination of data and information processing capacities of information technologies, and the creative and innovative capacity of human beings". From these two contrasting definitions, one can view KM as an extension of traditional information management whilst focussing on the synergistic outcome of combining information management and human creativity. Knowledge synchronisation involves having every new insight and every new piece of knowledge becoming instantly available to every employee across the organisation.

Zimbabwean organisations have been implementing knowledge management initiatives. Knowledge acquisition is defined as the transfer and transformation of problem-solving expertise from some knowledge source to a program (Buchanan et al, 1983). The main usable resources of knowledge are experts, textbooks, data bases, and experience from humans. Knowledge elicitation is a special kind of knowledge acquisition where the source of information is the human expert and a knowledge engineer. The knowledge engineer then interprets the data by abstraction into types and structures of knowledge, or simply modelling. Learning is defined in a broad sense as the acquisition of new skills and knowledge that results in changed behavior (Snyder & Cummings, 1998). Knowledge resides in, and is created by, both individuals and collectives. Knowledge management activities occur in building databases, measuring intellectual capital, establishing corporate libraries, building intranets, sharing best practices, installing groupware, leading training programs, leading cultural change, fostering collaboration, creating virtual organizations, etc. Absorptive capacity refers to the ability to process new knowledge as a function of an existing knowledge base. The primary purpose of knowledge management is to benefit the organisation and ultimately the customer. The use of information technology to exploit knowledge management has become a topical area. Knowledge embedded in products and services has been recognized as a primary source of sustainable competitive advantage in a knowledge-based economy.

Visible leadership and commitment of top management are prerequisites for successful knowledge management. There are six core processes of knowledge management that are all closely related and whose building blocks are illustrated by the figure 4 below (after Schulteis R., 1998):

**The Building Blocks of Knowledge Management**

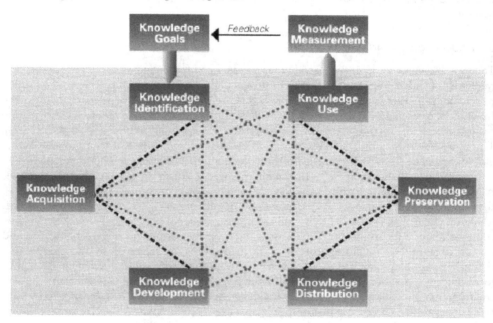

Fig. 4. Knowledge Management Building Blocks

Knowledge Management can be used to attain a cultural shift. Culture is an important contextual variable in the shaping and success of knowledge management, and knowledge sharing in particular. The New Generation of Knowledge Management is premised on the following perspectives:

a. *Tactical perspective* considers exploiting knowledge processes to achieve more effective enterprise operations, e.g. cost-reducing KM-supported innovations.
b. *Operational perspective* – attends to creating and fostering general KM practice and initiating and managing individual knowledge processes e.g. implementing lessons learnt programs.
c. *Knowledge implementation, manipulation and application perspective* - focuses on manipulating and applying knowledge to reflect how people and organisations deal with knowledge.

Knowledge can exist in either the explicit or tacit knowledge format. Explicit knowledge is structured formal knowledge, easily codified and articulated, for example a computer programme. Tacit knowledge is less structured, difficult to codify, articulate and share, and requires basic knowledge, experience and expertise as its basis. Management and coordination of diverse technology architectures, data architectures, and system architectures poses obvious knowledge management challenges. The focus of the technology-push model is on mechanistic information processing while the strategy-pull model facilitates organic sense making (Malhotra, 2005). The missing link between technologies and business performance is often attributable to the choice of technologies intended to fix broken processes, business models, or organizational cultures.

**Knowledge Management** (KM) refers to practices used by organizations to find, create, and distribute knowledge for reuse, awareness, and learning across the organization. For highly successful, adaptive and modern organisations, these organizations should have a clearly-enunciated and well-developed strategic information system plan (SISP), that consists of a strategy for both information planning and management, including the use of functions, and the salient features of information technology (IT) (Galliers, Swatman and Swatman, 1995).

### 3.3 ICTs for development (ICT4D)

The emergence and convergence of information and communication technologies (ICTs) has remained at the centre of global socio-economic transformations. ICTs include a wide range of services, applications and technologies, using various types of equipment and software, often running over telecommunications networks. ICTs have fundamentally changed the way people live, work, and interact socially in developing countries like Zimbabwe. There is a growing need to evaluate the social and economic impacts of ICTs and to create opportunities for capacity building that will ensure their beneficial use and absorption within national economies and civil society. ICTs do in fact have an impact on the standards of living and on poverty alleviation at various community levels, hence the focus on poverty reduction and development in Zimbabwe. ICT investment fosters higher long-term economic growth. The potential impact that ICT can have on individuals, businesses, and governments depends largely on how policies are formulated and technology and markets evolve.

As a result of the convergence of information, telecommunications, broadcasting and computers, the Information and Communication Technologies (ICTs) sector now embraces a large range of industries and services. The potential of ICTs to transform development is now receiving greater attention worldwide. There are many opportunities for developing countries like Zimbabwe to advance development through the innovative use of ICT. Innovations diffuse through a social system explained by the diffusion of innovation theory (Rogers, 2003). Knowledge-based activities have become increasingly important and pervasive worldwide, and ICT is a key foundation. Access to and development of ICT resources are increasingly recognised as crucial instigators of economic and social development. If ICTs are appropriately deployed, they have the potential to combat rural and urban poverty and foster sustainable development (Samiullah & Rao, 2000). Generally, investment in the development of the ICT infrastructure will result in improved economic efficiency and competitiveness; more efficient and effective education; healthcare and public administration; opportunities to exploit low factor costs in international markets; opportunities to increase social capital; and opportunities to bypass failing domestic institutions.

### 3.4 Sub-themes for knowledge frontiers

Knowledge frontiers for the major research problem can be the solution for sustainable growth and development, presented below as sub-themes.

### 3.4.1 Indigenous knowledge systems

The ever-growing process of the division of labour is producing ever more specialized knowledge, understandable only by specialists and experts. Specialized knowledge tends to be organized in egotistical cliques with monopolistic powers and privileges. Integrative knowledge keeps AKSs from disintegrating, but requires leadership and vision, which need to be improved throughout AKSs in institutional policy-making, implementation, monitoring and in integration of science, technology, innovation and knowledge policies with other development policies[3] (Jamel, 2004).

The African Indigenous Knowledge Systems (AIKS) and their contribution to knowledge frontiers for sustainable development are very important. Of interest in this context is the need to understand the importance and relevance of Indigenous knowledge systems and their link to sustainable growth and development in Africa and the world. It is also envisioned that priority be given to the significant contributions of Indigenous Knowledge Systems to knowledge creation, sustainable growth and development.

### 3.4.2 Transformational leadership

Leadership is "the ability to influence, motivate, and enable others to contribute toward the effectiveness and success of the organizations of which they are members." (House, 2004). The sub-theme covers theories, models and components of transformational leadership as well as demonstrates how the leadership style influence sustainable growth and development. Our interest is on how transformational leadership is viewed and its place in extending knowledge frontiers for sustainable growth and development. Since there are different types of transformational leadership, the paper focuses on the types that have an impact on knowledge frontiers for sustainable growth and development.

Leadership has two main forms:

- *Transactional leadership* involves exchanging rewards for services rendered and has the following attributes:
  - hierarchical
  - senior management decision making
  - task and reward focused
  - disempowering strategies
  - reduced creativity
- Transformational leaders transform the organization by developing vision, building commitment, and empowering followers. Transformational leadership is characterized by the following key attributes:
  - team building
  - shared vision of staff and leaders

---

[3] Science, technology and innovation constitute an important pillar of AKSs.

- staff involved in shared decision making
- feedback loop
- empowering strategies
- clear expectations and accountability of all

### 3.4.3 Open and distance learning and graduate employment

E-learning is learning supported or enhanced through the application of Information and Communications Technology (ICT), and has become an important pillar in open and distance learning. E-Learning can cover a spectrum of activities from supported learning to blended learning and to learning that is delivered entirely online. The Zimbabwe Open University (ZOU) is a State University, which was established on 1st March 1999 as an open and distance learning (ODL) university. The ZOU provides ODL throughout the ten (10) Regional Centres in each Province countrywide, as shown on the diagram below on Figure 5, and 1 Virtual Region for coordination and management of the e-learning platform accessible from anywhere in the world.

# ZOU Regional Centres

Fig. 5. The Regional Centres of Zimbabwe Open University

The ZOU offers pre-service and in-service ODL certificate, diploma and degree programmes. As an ODL university, ZOU is *accessible* to students from remote districts in all provinces of the country and reduces urban bias, provides easy and convenient accessibility to higher education, and there are no emigration to far countries.

### 3.4.4 Ethics and values

The essence of ethics is in addressing morally questionable acts and other dynamic relationships that affect ethical decision making. Morality is a first-order set of beliefs and practices about how to live a good life, whilst ethics is a second-order, conscious reflection on the adequacy of our moral beliefs. The essence of ethics takes note of the following observations and strategy:

- Codes of ethics do not necessarily lead to ethical behaviour.
- The core values we profess are not necessarily those by which we live.
- There is a place for compassion in leadership.
- Bureaucracy can come in conflict with ethics.
- Managers who vent their frustration on subordinates (who can do little about it) are not acting ethically.
- Study the ethics of organizations which have reputations for being ethical.
- Build ethics into organization policies and practices.
- Make sure quality and service and integrity permeate the entire organization.
- Develop high expectations of all members of your organization.

### 3.4.5 Emerging technologies

Current developments and debates on emerging technologies and their impact on knowledge creation, management and sustainable growth and development are consolidated. Trends and emerging technologies, innovations in various fields, and how emerging technologies are linked to knowledge frontiers for sustainable growth and development, is of paramount interest to this research problem, and can potentially make significant contributions to sustainable growth and development. The unceasing computing trends now focus on seamless integration, ubiquitous connectivity, embedded intelligence, user interface, and speed and capacity of computers. The notable key technology trends are in the following areas:

- Computing Hardware
- Networking and Wireless
- Knowledge Management
- Collaborative Software
- Security and Disaster Recovery

### 3.4.6 Entrepreneurship and community engagement

The main interest is on how the concepts of entrepreneurship and community engagement are impacting on expansion of knowledge and the sustainability of the systems. The paper is interested in seeking understanding on the perceptions of communities on academic institutions' role on knowledge, creation, development and dissemination.

### 3.4.7 Intellectual property and heritage

The links between intellectual rights and sustainable growth and development are important. The interest for the paper is on how to link intellectual property to challenges in intellectual property rights and impact on sustainable growth and development. Other related areas include intellectual property rights as a form of property rights, patents, and control over intangible and tangible assets. Intellectual property rules fundamentally affect the quality and availability of innovative ideas and products, and are therefore extremely important in achieving sustainable growth and development.

## 4. Methodology

Generally, the research design covers research philosophy, approaches, strategies, time horizons and methods. Research design involves the theory, conceptualization, formalization, operationalization of variables, choice of methods, data collection techniques, and data analysis (Deflem, 1998). The research is a case study for Zimbabwe. The methodology used was largely *qualitative on human capital development and knowledge management*, and also quantitative on Infodensity for Zimbabwe. The qualitative approach was focused on human capital development and knowledge frontiers in Zimbabwe. Data on Infodensity was obtained from the International Telecommunications Union (ITU, 2011).

The *qualitative research* was used to deepen our understanding of the human capital development issues and knowledge frontiers, and its link to economic growth. For this research external secondary data was largely used, for it saves on time and reduces data gathering costs. The following data collection techniques were used in this case study:

- Formal meetings and focus group discussions
- Face-to-face oral structured interviews
- Questionnaires on knowledge frontiers
- Secondary data and records observation

The face-to-face interviews allowed for in-depth knowledge sharing, helped to develop the bigger picture on knowledge frontiers and was good for networking. Focus group discussions were held with selected regulatory, training and research institutions to pick up grassroots input and in developing ideas, whist sharing latent knowledge spontaneously.

## 5. Analysis of results

### 5.1 Infodensity for Zimbabwe

Data on infodensity was obtained from the International Telecommunications Union (ITU, 2011, http://www.itu.int/en/publications/), and the analysis is shown below on figures 6. Zimbabwe has about 10% internet penetration rate. The number of internet users per 100 inhabitants (%) in Zimbabwe is benchmarked against other 18 countries in East and Southern Africa for the period 2000-2010.

The mobile density of Zimbabwe more than doubled from 24% in 2009 to 60% in 2010. The number of mobile users for Zimbabwe is benchmarked against other 18 neighbouring countries in Africa for the period 2000-2010 and is shown on Figure 7 below. Botswana has the highest mobile density followed by South Africa and Mauritius.

The mobile density for Zimbabwe alone is shown on the Figure 8 below. The mobile density for Zimbabwe has risen astronomically between 2008 and 2010, and has one of the highest growth rate for mobile density among the 18 countries.

From the above chart on Figure 8, the mobile density for Zimbabwe has risen astronomically between 2008 and 2010, and has one of the highest growth rate for mobile density among the 18 countries in the region of Africa.

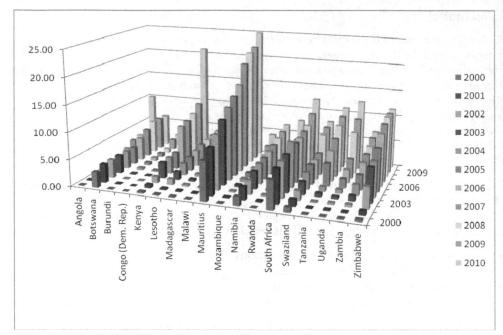

Fig. 6. Number of internet users for Zimbabwe benchmarked against others neighbouring African countries for the period 2000-2010

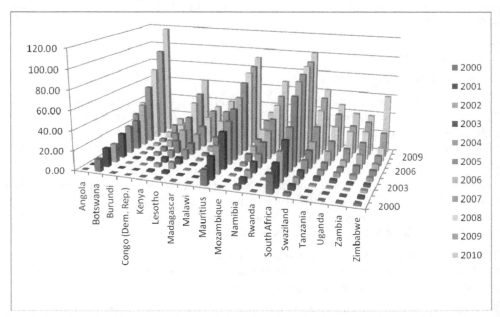

Fig. 7. Number of mobile users for Zimbabwe benchmarked against other neighbouring African countries for the period 2000-2010

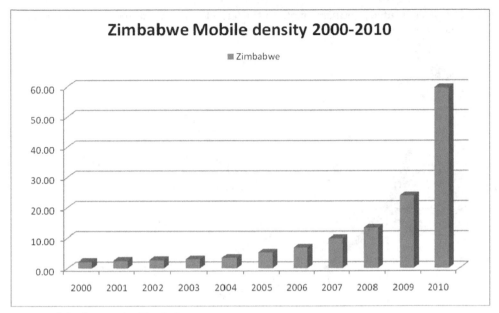

Fig. 8. Mobile density for Zimbabwe

## 5.2 Sustainable development and human capital

Human capital development is central to capacity development. Capacity development looks at the overall system, environment or context within which individuals, organisations and societies operate and interact. *Human capital is* about "the knowledge, skills and competences and other attributes embodied in individuals that are relevant to economic activity" (OECD, 1998). If education for sustainable development (ESD) is to become a central aspect of all education, supporting structures at the global, national and local levels are required. Investment in human capital and technology is the sustainable long-term solution to the above problem and its symptoms. Human capital is an asset and a factor of production, that can be measured at individual, corporate and national levels. The ICT revolution, at institutional and regional collaboration levels, requires extensive investments into people (labour) and capital for the infrastructure and equipment (Kabanda, 2008). The Cobb-Douglas production function relates the revolutionary technological change or productivity levels from ICT to labour and capital. A Cobb-Douglas production function of the form:

$$Q = A\ K^a\ L^b$$

is used for the analysis of technological progress and attended economic growth, where A, a and b are empirical parameters. In this context, the key parameters driving productivity include:

- $K$ = capital input (very meaningful mounts)
- $L$ = labour input (high technical competence and human capital)

Sufficient investment in human capital and technology increases the factor of production by numerous multiples, and consequently reduces the unit cost of production of items, e.g.

assembly of computers, printers, motor vehicles, etc., as illustrated in Figure 9 below (Kabanda, 2008):

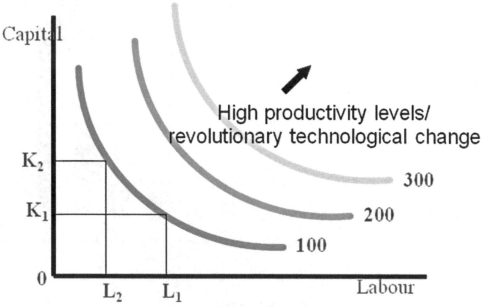

Fig. 9. High productivity levels due to extensive investment in human capital and technology

The unit cost in production decreases as a result of extensive investments in both technology for large throughput and human capital development.

Duration of schooling and levels of qualification are the standard measures used in human capital development, but these are far from capturing the extent of human capital. The major human capital development challenges faced in Zimbabwe in both education and technology development during the period 2005-2010 include the following challenges to:

1. Improve access, equity and retention
2. Improve quality assurance
3. Increase flexibility and responsiveness to meet market demand and the priorities of the national development agenda
4. Increase diversity in programmes and forms of delivery
5. Increase efficiency & effectiveness by rationalizing existing resources, improving management systems, and diversifying sources of finance.
6. Corporate Governance, e.g. regulatory mechanisms for policy implementation, facilitating regional integration and international cooperation, etc.
7. National level Institutional challenges:
   • Inadequate funding and inappropriate funding mechanisms
   • Maintenance of infrastructure, equipment, ICTs and facilities
   • Leadership development & planning capacity

8. General national Human Capital challenges identified across all sectors, e.g. brain drain, conditions of service and innovation, addressing gender and societal imbalances, curricular reforms, capacity building and re-skilling.

The current and future position of education for sustainable development (ESD) in Zimbabwe is guided by the following logical model that consolidates the Zimbabwean experiences during the period 2005-2010, and which model has the following 4 pillars shown on Figure 10.

# ESD LOGICAL FRAMEWORK

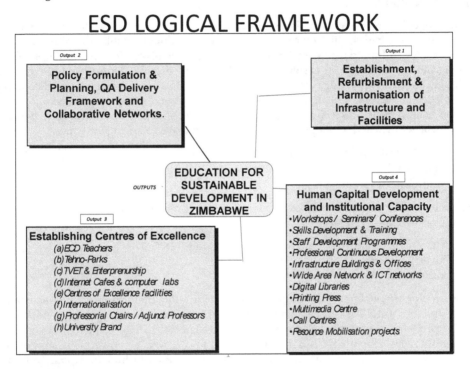

Fig. 10. The ESD Logical Framework for Zimbabwe

ESD in Zimbabwe provides a balance between human capital and social capital. Emphasis on social relationships *counteracts excessive individualisation* and *longer-term perspective* into policy-making. In the Zimbabwean context of ESD, social capital reintroduces a *moral dimension* into educational and economic thinking. The solution to the human capital development challenges in Zimbabwe is a long-term sustainable strategy. In 2011, the University of Phoenix Research Institute produced a 19-page publication (http://cdn.theatlantic.com/static/front//docs/sponsored/phoenix/future_work_skills_2 020.pdf.) which indicated the following ten skills for the future workforce:

- Sense-making
- Social Intelligence
- Novel and adaptive thinking
- Cross-cultural competency

- Computational thinking
- New-media literacy
- Transdisciplinarity
- Design mindset
- Cognitive load management
- Virtual collaboration.

## 6. Conclusions

Knowledge-driven sustainable development requires relevant and efficient development knowledge. Zimbabwe, like other African countries, should establish African Knowledge Systems (AKSs) from which knowledge policies may precipitate. To some extent, knowledge for sustainable development is something knowable, modelable, reformable and manageable through effective knowledge policies (Hamel, 204). However, sufficient understanding of knowledge in the context of an African knowledge corpus and AKSs brings some completeness in the achievement of sustainable growth and development. The knowledge frontiers for sustainable growth and development are categorised into the following three areas:

- Knowledge for modernization and development
- Indigenous and traditional knowledge, and
- Faith-based knowledge.

Zimbabwe is faced with the following three major sustainable development problems that require knowledge frontiers as a solution:

1. Attainment of the Millenium Development Goals (MDGs)
2. Poverty reduction
3. Social problems value chain

This exciting wave of new knowledge provides knowledge opportunities for the sustainable development of the African continent, perhaps through adequate knowledge policies. The major problem of under-development characterised by the huge challenge to achieve the Millennium Development Goals (MDGs) is on knowledge empowerment supported by Information and Communication Technologies (ICTs). Rich in human capital and natural resources, Zimbabwe can achieve sustainable growth and development.

The perspective on knowledge as a sustainable resource for development indicates that knowledge is a necessary condition for sustainable development in various constructed environments. Economic environments require full access to modern and efficient development knowledge. Knowledge can also be viewed alongside a continuum: data – information – knowledge – wisdom. Development knowledge refers broadly to the totality of representative mental or abstract structures and constructions related to sustainable development. Future economic competiveness is premised on knowledge management. Zimbabwean organisations have been implementing knowledge management initiatives. Knowledge elicitation is a special kind of knowledge acquisition where the source of information is the human expert and a knowledge engineer. The use of information technology to exploit knowledge management has become a topical area.

Knowledge frontiers for the major research problem can be the solution for sustainable growth and development, presented as sub-themes.

1.  **Indigenous Knowledge Systems**
    - African Knowledge Systems (AKSs) &
    - African Indigenous Knowledge Systems (AIKS)
2.  **Transformational Leadership**
    - *Transactional leadership vs Transformational leadership*
3.  **Open and Distance Learning and Graduate Employment**
4.  **Ethics and values**
5.  **Emerging Technologies**
6.  **Entrepreneurship and Community Engagement**
7.  **Intellectual Property and Heritage**

The way forward for sustainable development in Zimbabwe is on:

- Promotion of technology development, transfer and diffusion to Africa and further develop technology and knowledge available in African centres of excellence;
- Supporting African efforts to develop affordable transport systems and infrastructure that promote sustainable development and connectivity in Africa;
- Leadership development & planning capacity;
- Enhancing science, technology and enterprise (SMEs) development;
- Harmonisation & development of Private Public Partnerships (PPPs), and
- Utilisation of Human Capital.

## 7. References

Bergeron, B. (2003). Essentials of knowledge management . New Jersey: John Wiley & Sons Inc. Hoboken,

Buchanan, Hayes-Roth F., Waterman D.A. and Lenat D. (1983), Building Expert systems, p127-159, in *Constructing an Expert System by Buchanan: Reading, Massachusetts:* Addison Wesley *Publishing Company, 1983.*

Deflem, M. (1998): 'An Introduction to Research Design' [www.mathieudeflem.net] (10/12/08)

Galliers R.D, Swatman P.M.C And Swatman P.A (1995): "Strategic Information Systems Planning: Deriving comparative advantage from ED", Journal of Information Technology, vol. 10, pp 149-157

Hamel J.L. (2004), "Knowledge Policies for Sustainable Development in Africa: A strategic framework for good governance", Economic Commission for Africa (ECA)/SDD Working Paper.
http://www.uneca.org/estnet/ecadocuments/knowledge_policies_for_sustainabl e_development_in_Africa.

Heeks, R. (1999) 'Information and Communication Technologies, Poverty and Development'. Development Informatics Working Paper Series, Paper No. 5, June 1999, IDPM, Manchester.
http://www.man.ac.uk/idpm/idpm_dp.htm#devinf_wp.

House, R. J. (2004). *Culture, Leadership, and Organizations: The GLOBE Study of 62 Societies,* SAGE Publications, Thousand Oaks.

International Telecommunications Union (2011), World Telecommunication/ICT Indicators Database: ITU, April, 2011, http://www.itu.int/ITU-D/ict/statistics/index.html.

Kabanda, G. (2008), " *Collaborative opportunities for ICTs development in a challenged African environment"*, published in the Journal of Technology Management & Innovation, August 2008, Volume 3 Number 3 of 2008, pages 91-99, ISSN 0718-2724.

Kabanda, G. (2011), *"Impact of information and communication technologies (ICTs) on millennium development goals (MDGs): Context for diffusion and adoption of ICT innovations in East and Southern Africa"*, Journal of African Studies and Development, August 2011 Volume 3 (8), paper JASD-10-038, Available online http://www.academicjournlas.org/ JASD, ISSN – 2141 -2189 ©2011 Academic Journals.

Kekana, N. (2002) Information Communication and Transformation: A South African Perspective. Communication, Vol. 28 No.2, p54-61

Kurasha, J. (2011), "Digital Natives", Public lecture of Open University of Tanzania on 11th July, 2011, 3rd African Council for Distance Education (ACDE) Conference & General Assembly proceedings, Open University of Tanzania 2011.

Malhotra, Y. (1998). 'Deciphering the Knowledge Management Hype' The Journal for Quality & Participation, Association for Quality & Participation.

Malhotra, Y. (2005). 'Integrating knowledge management technologies in organizational business processes: getting real time enterprises to deliver real business performance'. Journal Of Knowledge Management. Volume 9 (1)

OECD (1998) *Human Capital Investment: An International Comparison*, Paris, Organisation for Economic Cooperation and Development.

Rogers, E.M. (2003). Diffusion of Innovations (5th Edition). New York. The Free Press.

Samiullah, Y and S. Rao (2000) 'Role of ICTs in Urban and Rural Poverty Reduction'. Paper prepared for MoEF-TERI-UNEP Regional Workshop for Asia and Pacific on ICT and Environment, May 2000, Delhi.
http://www.teri.res.in/icteap/present/session4/sami.doc

Sciadas G. (2003), "Monitoring the Digital Divide and beyond", The Orbicom project publication, in collaboration with the Canadian International Development Agency, the *InfoDev* Programme of the World Bank and UNESCO: Claude-Yves Charron, Canada, 2003 (ISBN 2-922651-03-7), page 10. *http://www.orbicom.uqam.ca*.

Snyder, W. & Cummings, T H. (1998). 'Organization learning disorders: Conceptual model and intervention hypothesis'. Human Relations, Volume 51(7), p873-895.

UNCTAD (2006), The United Nations Conference on Trade and Development "The least developed countries report for 2006: developing productive capacities", Paper number UNCTAD/LDC/2006, United Nations New York and Geneva, 2006 (http://www.unctad.org/en/docs/ldc2006_en.pdf).

UNESCO (2010), "Universities and the Millennium Development Goals Report 2007", Association of Universities and Colleges in Canada, WDI publication, presented at the Association of Commonwealth Universities conference, Cape Town, April 2010.

Watson, I. (2003). Applying knowledge management techniques for building corporate memories. Elsevier Science (USA). [Publisher unknown]

World Bank (2002), 'Constructing Knowledge Societies: New Changes for Tertiary Education', Education World Bank

World Bank (2002). *ICT Sector Strategy Paper*. World Bank.
http://info.worldbank.org/ict/ICT_ssp.html

World Bank (2003). *Lifelong learning in the global knowledge economy: Challenges for developing countries*. Washington, DC: World Bank.

World Development Indicators (2004). "E-Strategies: Monitoring and Evaluation Toolkit", The World Bank INF/GICT Volume 6.1B 3rd January, 2005.
http://www.worldbank.org/ict/.

# Permissions

The contributors of this book come from diverse backgrounds, making this book a truly international effort. This book will bring forth new frontiers with its revolutionizing research information and detailed analysis of the nascent developments around the world.

We would like to thank Prof. Aurora A.C. Teixeira, for lending his expertise to make the book truly unique. He has played a crucial role in the development of this book. Without his invaluable contribution this book wouldn't have been possible. He has made vital efforts to compile up to date information on the varied aspects of this subject to make this book a valuable addition to the collection of many professionals and students.

This book was conceptualized with the vision of imparting up-to-date information and advanced data in this field. To ensure the same, a matchless editorial board was set up. Every individual on the board went through rigorous rounds of assessment to prove their worth. After which they invested a large part of their time researching and compiling the most relevant data for our readers. Conferences and sessions were held from time to time between the editorial board and the contributing authors to present the data in the most comprehensible form. The editorial team has worked tirelessly to provide valuable and valid information to help people across the globe.

Every chapter published in this book has been scrutinized by our experts. Their significance has been extensively debated. The topics covered herein carry significant findings which will fuel the growth of the discipline. They may even be implemented as practical applications or may be referred to as a beginning point for another development. Chapters in this book were first published by InTech; hereby published with permission under the Creative Commons Attribution License or equivalent.

The editorial board has been involved in producing this book since its inception. They have spent rigorous hours researching and exploring the diverse topics which have resulted in the successful publishing of this book. They have passed on their knowledge of decades through this book. To expedite this challenging task, the publisher supported the team at every step. A small team of assistant editors was also appointed to further simplify the editing procedure and attain best results for the readers.

Our editorial team has been hand-picked from every corner of the world. Their multi-ethnicity adds dynamic inputs to the discussions which result in innovative outcomes. These outcomes are then further discussed with the researchers and contributors who give their valuable feedback and opinion regarding the same. The feedback is then collaborated with the researches and they are edited in a comprehensive manner to aid the understanding of the subject.

Apart from the editorial board, the designing team has also invested a significant amount of their time in understanding the subject and creating the most relevant covers. They scrutinized every image to scout for the most suitable representation of the subject and create an appropriate cover for the book.

The publishing team has been involved in this book since its early stages. They were actively engaged in every process, be it collecting the data, connecting with the contributors or procuring relevant information. The team has been an ardent support to the editorial, designing and production team. Their endless efforts to recruit the best for this project, has resulted in the accomplishment of this book. They are a veteran in the field of academics and their pool of knowledge is as vast as their experience in printing. Their expertise and guidance has proved useful at every step. Their uncompromising quality standards have made this book an exceptional effort. Their encouragement from time to time has been an inspiration for everyone.

The publisher and the editorial board hope that this book will prove to be a valuable piece of knowledge for researchers, students, practitioners and scholars across the globe.

# List of Contributors

**Musa Jega Ibrahim**
Economic Research and Policy Department, Islamic Development Bank, Jeddah, Kingdom of Saudi Arabia

**Mónica L. Azevedo, Sandra T. Silva and Óscar Afonso**
CEF.UP, Faculty of Economics, University of Porto, Porto, Portugal

**Tony Smith**
Iowa State University, USA

**Aurora A.C. Teixeira**
CEF.UP, Faculty of Economics, University of Porto, Porto, INESC Porto, OBEGEF, Portugal

**Byoung Soo Kim**
Korea Institute of S&T Evaluation and Planning, Republic of Korea (South Korea)

**Steven R. Walk**
Old Dominion University, USA

**António C. Moreira**
DEGEI, GOVCOPP, University of Aveiro, Portugal

**Ana Carolina Carvalho**
ISCA, University of Aveiro, Portugal

**Oscar Afonso, Alexandre Almeida and Cristina Santos**
CEF.UP, Faculty of Economics, University of Porto, Porto, Portugal

**Özlem Olgu**
College of Administrative Sciences and Economics, Koc University, Rumeli Feneri Yolu, Sariyer, Istanbul, Turkey

**Gabriel Kabanda**
Zimbabwe Open University, Harare, Zimbabwe

9 781632 384355